### *Study Guide to accompany*

# INTRODUCTION TO General, Organic & Biochemistry

*Third Edition*

***Bettelheim & March***

***William Scovell***

**Bowling Green State University**

Saunders College Publishing
**Harcourt Brace Jovanovich College Publishers**
Fort Worth Philadelphia San Diego
New York Orlando Austin San Antonio
Toronto Montreal London Sydney Tokyo

Printed in the United States of America.

Bettelheim/March: Study Guide to accompany INTRODUCTION TO GENERAL ORGANIC AND BIOCHEMISTRY, 3/E

ISBN # 0-03-054678-8

23 095 9876543

# TO THE STUDENT

I hear, and I forget
I see, and I remember
I do, and I understand

"Ancient Chinese Proverb"

The aim of this Study Guide is simply to aid the student in increasing understanding of the material presented in Introduction to General, Organic and Biochemistry by Bettelheim and March, and to make studying more efficient. The general philosophy or approach put forth here should not only be useful in this course, but in many courses in which the learning of specific new concepts and the development of problem solving abilities is important. The initial focus of attention should be to determine the immediate objectives or aims of each chapter. Therefore study objectives will begin every chapter in order to point out aspects worthy of special emphasis in your studies.

Once the major objectives have been read over carefully, one must ask, "How best can I study to develop a firm understanding and command of the new vocabulary, concepts and problem solving techniques associated with the material?" The approach to this end will be to define four study practices.

1. A list of key terms follows the summary at the end of each chapter in the text. Additional terms of importance from the text will be listed in the Study Guide either individually, or in cases in which two or more terms are related, they will be listed together. This latter grouping will serve to emphasize that the similarities and differences between these terms should be clearly understood. The terms are basic to the understanding of the new material and will provide the very foundation of your understanding. Review them together with the objectives before reading the chapter. After reading the text, reread the objectives and define each term. To deepen your understanding of the terms, include a specific example in each case where this is appropriate. Refer to the Glossary in the text to provide clear definitions to all the terms. **Continually review** the definitions before each study session or until the responses are clear and spontaneous.

2. In each chapter, a few concepts provide the basis for grasping many of the sometimes apparently diverse, but interrelated subjects presented. In some chapters, a selected number of concepts will be elaborated on in order to "pull together" some of the ideas and to deepen understanding of the material.

3. An integral part of chemistry involves the solving of both mathematical and conceptual problems. The general thought processes and interrelationships helpful in reaching these solutions will be briefly outlined. These problems will, in some cases, include the detailed solutions of some of the problems in the text.

4. At the point which the student feels confident that the material has been mastered, a SELF-TEST will provide a means for evaluating the progress toward reaching the major objectives in the chapter. Answers to all the questions and problems will be included, with explanations accompanying some of the answers.

The last section in the study guide will present the answers to the even-numbered, end-of-chapter problems in the text.

## SOME OBSERVATIONS AND A SUGGESTED STUDY APPROACH

1. Typically students who obtain above average grades will read the chapter a minimum of twice, if not three times. I point it out only because I find many freshman are surprised to learn that this effort is actually the **norm** and to be sure, not unusual.

2. Reading the chapter prior to the material being discussed in class usually affords the student the unique advantage of being able to clarify questions on the material at the time the professor is introducing it to the class. Unfortunately, for many students, this often becomes very difficult, if not impossible. However, try it. You may find it so valuable that it may become a part of your personal study strategies.

3. Take good notes in class and as soon after class as possible, **recopy the notes into another notebook in a more organized fashion**. Do this at the same time that you refer to your textbook and annotate your class notes with extra definitions or comments from the text. I have found that this has been especially helpful to many students.

4. Knowledge of a subject begins with **memorizing** the meaning of new terms, concepts and equations. However, **understanding** the material goes far beyond the memorization of these elements in a "vacuum." It requires that the interrelationships between all these elements be clearly resolved. This requires a continual review of the material and a major input of time and effort into solving problems and answering focused questions. Therefore, it is important that you commit yourself to working as many problems as possible. There is no substitute for this.

I would like to outline a study procedure that many students have found helpful. This, of course, will not be for everyone. Each student must develop routines which serve him or her best. Try this approach if you wish. If it helps, use it.

1. Review the chapter objectives and important new terms in the study guide. This will alert you to the key ideas and concepts and help to underline their importance before your reading.

2. Read the (text) chapter with these points in mind.

3. Record good notes in class and as soon after class as possible, rewrite the notes. Clarify all questionable points in the lecture notes **AT THIS TIME**, with the aid of your text. For those concepts which are not clear, ask the professor immediately after the next lecture or bring up this point at his or her next office hour.

4. Write out the definitions of all important new terms.

5. Review any worked problems in the chapter, with the aim of defining the general thought processes involved in solving the problem.

6. Now, the assigned questions and problems at the end of each chapter in the text should be worked and the material in the Study Guide should be studied. The guide will be most useful in cases where some additional explanation or worked examples are helpful. In some cases, simply rephrasing of text material may bring out the meaning more clearly.

7. Reread the chapter, the rewritten class notes, chapter objectives, important terms and equations and the focused review section.

8. Return to doing questions and problems that were initially unclear or difficult and then do additional ones.

9. Do the SELF-TEST in the study guide.

## TABLE OF CONTENTS

"It is only the first step
which takes the effort"
- Madame Marie Vichy-Deffand

# 1

# Matter, Energy and Measurements

## CHAPTER OBJECTIVES

After studying Chapter 1 and the material in Appendices A and B, and working the assigned exercises in the text and the study guide, you should be able to do the following.

1. Distinguish between a fact, a hypothesis and a theory. Indicate how they are associated with the scientific method.

2. Convert large and small numbers to an equivalent form in exponential notation. Carry out the reverse procedure.

3. Perform addition, subtraction, multiplication and division calculations using exponential notation. Evaluate the answers to insure that there are the proper number of significant figures.

4. List the word prefixes used in the metric system and the corresponding values.

5. Define temperature and indicate temperature scales in common use. Write the equations which relate temperature in (i) °C to °F; (ii) °F to °C (iii) °C to K; perform temperature conversions from one scale to another.

6. Using the factor-label method, perform conversions between the metric and English units, English to metric units, in addition to conversions within the metric system.

7. List the three states of matter and the characteristics which are unique to each state.

8. Distinguish between density and specific gravity.

9. Explain the basic difference between **ex**tensive and **in**tensive properties of a substance and give a few examples of each.

10. Define energy and list the six different forms of energy pointed out in this chapter.

11. State the Law of Conservation of Energy.

12. Define heat and specific heat and indicate the units of each.

13. Using the rules for determining the number of significant figures, evaluate the number of significant figures in numbers listed throughout the chapters.

14. Define the important terms and comparisons in this chapter and give specific examples where appropriate.

## IMPORTANT TERMS AND COMPARISONS

Matter and Energy
Fact, Hypothesis and Theory
Coefficient and Exponent
Kilo- and Milli-
Mega- and Micro
Centi-
Mass and Weight
Centigrade, Fahrenheit and Kelvin Scale
States of Matter
Density and Specific Gravity
Intensive and Extensive Properties
Potential and Kinetic Energy
Law of Conservation of Energy
Temperature and Heat
Calorie and Joule

## FOCUSED REVIEW OF CONCEPTS

a. Interrelationships

(i) $°F = 9/5(°C) + 32$

(ii) $°C = 5/9(°F-32)$

(iii) $K = °C + 273$

(iv) Density = mass/volume; $D = M/V$
D is a **physical quantity** and has **units**

(v) Specific Gravity (S.G.) =

$$\frac{\text{Density of substance of interest}}{\text{Density of water}}$$

$S.G. = D_A/D_{H_2O}$

S.G. is a unitless number

(vi) Heat (cal) = Specific heat (cal/g-deg) x mass (g) x $(T_2-T_1)$

It should be noted that both equations (i) and (ii) define the relationship between the temperature in degrees Centigrade and in degrees Fahrenheit. Therefore, you should be able to rearrange equation (i) to arrive at equation (ii).

Write eqn (i)

$°F = 9/5\ (°C) + 32$

Subtract 32 from each side

$$°F-32 = 9/5(°C)+32-32$$

$$(°F-32) = 9/5(°C)$$

Multiple each side by 5/9

$$5/9(°F-32) = (5/9)(9/5)(°C)$$

This is eqn (ii)

$$5/9(°F-32) = °C$$

Equation (iii) should be put to memory since many calculations to be performed in subsequent chapters (e.g., Chapter 5) will require that the temperature be expressed in degrees Kelvin (K).

Density is an intensive property of a substance and has units of mass/vol, which although usually expressed in g/mL, can be expressed in others units (i.e., lb/in$^3$, etc.). Recall also that **in**tensive properties are those which are **in**trinsic characteristics of the substance and therefore do not depend on the quantity of the substance.

Specific gravity is a ratio of the density of any substance to the density of water at 20°C. Therefore, specific gravity is simply a number, with no units.

## PROBLEM SOLVING METHODS (Refer to Appendix Material in Text)

A. Exponential Notation: It is important to be able readily to express large and small numbers in exponential or scientific notation. This requires that the number be written as a **coefficient,** with a value between 1 and 10, **multiplied** by 10 raised to some power. For example,

$1\underline{00} = 1 \times 10^2 = 1 \times (1\underline{00})$ — The solid line underlines the number of places the decimal point has been moved in each case.

$2\underline{0{,}000} = 2 \times 10^4 = 2 \times (1\underline{0{,}000})$

$5{,}\underline{200{,}000} = 5.2 \times 10^6 = 5.2 \times (1{,}\underline{000{,}000})$

Note in these examples that when the LARGE number (greater than 1) is changed to exponential notation form, the decimal point is moved to the LEFT by 2, 4 or 6 places, respectively. The positive exponent simply expresses the number of places that the decimal has been moved to the left.

$0.\underline{001} = 1 \times 10^{-3} = 1 \times (0.\underline{001})$

$0.\underline{00004}3 = 4.3 \times 10^{-5} = 4.3 \times (0.\underline{00001})$

$0.\underline{02}00100 = 2.001 \times 10^{-2} = 2.001 \times (0.\underline{01})$

Similarly, in changing these SMALL numbers (less than one) to exponential notation, the decimal point is moved to the RIGHT by 3, 5 or 2, respectively.

**a. MOVING THE DECIMAL POINT** in Numbers Already in Exponential Notation.

One of the most common difficulties that students encounter occurs when doing multiple step calculations. Midway through the calculation, numbers such as the following may appear.

$16{,}743 \times 10^{14}$ $\qquad$ $0.00713 \times 10^{6}$ $\qquad$ $0.0000431 \times 10^{-6}$

To change these numbers to proper form (i.e., the coefficient being between 1 and 10), the decimal point must be moved. Which way is always the question!

**The rule to remember is:**

IF THE **COEFFICIENT IS DECREASED** in magnitude, the **EXPONENT MUST BE INCREASED ACCORDINGLY.**

**IF THE COEFFICIENT IS INCREASED,** the **EXPONENT IS DECREASED.**

For example:

$1\underline{6{,}743} \times 10^{14} \rightarrow 1.6743 \times 10^{18}$

**coefficient decreased**, decimal moved **4** places
**exponent increased by 4**

$0.\underline{007}13 \times 10^{6} \rightarrow 7.13 \times 10^{3}$

**coefficient increased,** decimal moved **3** places
**exponent decreased by 3**

$0.\underline{00004}31 \times 10^{-6} \rightarrow 4.31 \times 10^{-11}$

**coefficient increased**, decimal moved **5** places
**exponent decreased by 5**

Exercise I: Carry out these manipulations to develop skill in these conversions.

| Number | Exponential Notation (with coefficient between 1 and 10) |
|---|---|
| 1. $0.00102 \times 10^{1}$ | 1,02 x 10^-2 |
| 2. $632 \times 10^{-14}$ | 6,32 x 10^-12 |
| 3. $0.00700 \times 10^{43}$ | 7 x 10^40 |
| 4. $13{,}642{,}000 \times 10^{-1}$ | 1.3,642 x 10^46 |
| 5. $11 \times 10^{23}$ | 1.1 x 10^24 |
| 6. $0.0000006 \times 10^{-14}$ | 6 x 10^-21 |
| 7. $137 \times 10^{-10}$ | 1,37 x 10^-8 |

8. $0.006 \times 10^{0}$ ____________________

9. $7{,}423{,}000{,}000 \times 10^{17}$ ____________________

10. $0.4361 \times 10^{31}$ ____________________

In the frequently used manipulations with numbers in scientific notation outlined below, a TWO-STEP procedure can be used in each case.

**b. ADDITION OR SUBTRACTION** in Exponential Notation

Procedure: (i) change the numbers so all have the same exponent, irrespect of the size of the coefficient
(ii) Add (or subtract) the coefficients; the exponent remains the same

Ex. a.

$$\begin{array}{l} 230 \times 10^{-8} \\ \underline{6.10 \times 10^{-5}} \end{array} \quad \xrightarrow{(i)} \quad \begin{array}{l} 23 \times 10^{-7} \\ \underline{610 \times 10^{-7}} \end{array} \quad \xrightarrow{(ii)} \quad 633 \times 10^{-7} = 6.33\times10^{-5}$$

b.

$$\begin{array}{l} \phantom{-}340 \times 10^{6} \\ - \underline{220 \times 10^{5}} \end{array} \quad \xrightarrow{(ii)} \quad \begin{array}{l} \phantom{-}340 \times 10^{6} \\ - \underline{22 \times 10^{6}} \end{array} \quad \xrightarrow{(ii)} \quad 318 \times 10^{6} = 3.2\times10^{8}$$

**c. MULTIPLICATION** in Exponential Notation

Procedure: (i) Multiple the coefficients directly
(ii) Algebraically **add** the exponents

Ex. a. $(6.0 \times 10^{-14})(3.0 \times 10^{6})$ (i) → 18
(ii) → $\times\ 10^{-8}$
$\underline{18 \times 10^{-8}}$

$1.8 \times 10^{-7}$

b. $(1 \times 10^{-14})(7 \times 10^{-2})$ (i) → 7
(ii) → $\times\ 10^{-16}$
$\underline{7 \times 10^{-16}}$

**d. DIVISION** in Exponential Numbers

Procedure: (i) Divide the coefficients directly
(ii) Algebraically subtract the exponents

Ex. a. $\left(\frac{3 \times 10^{-7}}{1 \times 10^{7}}\right)$ (i) → 3

(ii) → $\times 10^{-7-7}$

$\underline{3 \quad \times 10^{-14}}$

b. $\left(\frac{3 \times 10^{7}}{1 \times 10^{-7}}\right)$ (i) → 3

(ii) → $\underline{\times 10^{7-(-7)}}$

$\underline{3 \quad \times 10^{14}}$

c. $\left(\frac{2.14 \times 10^{-3}}{8.60 \times 10^{+4}}\right)$ (i) → 0.249

(ii) → $\underline{\times 10^{-3-4}}$

$0.249 \times 10^{-7} = \underline{2.49 \times 10^{-8}}$

d. $\left(\frac{5.91 \times 10^{-4}}{1.61 \times 10^{-12}}\right)$ (i) → 3.67

(ii) → $\underline{\times 10^{-4+12}}$

$\underline{3.67 \quad \times 10^{8}}$

B. Unit Conversions

**(i)** All conversions within the metric system differ by some power of 10 as reflected in the prefixes. For example, centi = $10^{-2}$; milli = $10^{-3}$; micro = $10^{-6}$; nano = $10^{-9}$. These conversions are accomplished simply by moving the decimal point. Memorize these prefixes and those in Table 1.2 (text).

The strategy in these problems is to multiply the starting physical quantity (a number with a unit) by a conversion factor which will both change the number and also its units.

AT EACH STEP IN THE CONVERSION, ONE UNIT MUST CANCEL OUT. AT NO POINT IN THE PROCESS ARE THERE MORE UNITS THAN YOU START WITH. This conversion process is continued until the desired unit (or units) is (are) obtained. In each case, clearly understand what the initial units are and also what the units are you want to convert to.

Ex. 1 Change 2 ft to inches. Start with **feet;** Convert to **inches**

$$\left(\frac{2\ \cancel{ft}}{1}\right)\left(\frac{12\ in}{1\ \cancel{ft}}\right) \rightarrow \frac{24\ in}{1} = 24\ inches$$

starting physical quantity — conversion factor

Note that in this one-step conversion, 2 units cancel (ft) and only one unit (in., the desired unit) remains. If the correct conversion factor had been used, but incorrectly inverted, **no units would have canceled.** For example:

$$\left(\frac{2\ ft}{1}\right)\left(\frac{1\ ft}{12\ in}\right) \rightarrow \frac{2\ ft^2}{12\ in}$$

Units did not cancel. This incorrect set-up produced more units ($ft^2$/in) than in the starting quantity (ft). This is your signal that the conversion factor is inverted.

Ex. 2 Convert 3.2 milliliters to nanoliters.

Start with **milliliters;** want **nanoliters.** Depending on the conversion factors you know, this problem and many others, can be done in one or perhaps a few steps.

Route 1: $\left(\frac{3.2\ \not{mL}}{1}\right)\left(\frac{10^6\ nL}{1\ \not{mL}}\right) = 3.2 \times 10^6\ nL$

Route 2: $\left(\frac{3.2\ \not{mL}}{1}\right)\left(\frac{10^3\ \not{\mu L}}{1\ \not{mL}}\right)\left(\frac{10^3\ nL}{\not{\mu L}}\right) = 3.2 \times 10^6\ nL$

Although 1 nL = $10^{-9}$ L = $10^{-6}$ mL = $10^{-3}$ µL, you should note that the conversion factors are:

$\left(\frac{10^9\ nL}{L}\right), \left(\frac{10^6\ nL}{mL}\right), \left(\frac{10^3\ nL}{\mu L}\right)$

In each case, the exponents are **positive** to show the number of nanoliters contained in the larger volume unit.

Exercise II: Convert the quantities shown below into a metric equivalent. Since all the quantities are in the metric system, the conversion requires only moving the decimal to the left or the right.

| | | | | |
|---|---|---|---|---|
| 1. | 1.32 L | ______ mL | ______ µL | ______ nL |
| 2. | 6.32 mg | ______ g | ______ kg | ______ Mg |
| 3. | 0.039 cm | ______ m | ______ km | ______ µm |
| 4. | 0.36 µs | ______ ms | ______ s | ______ ns |

**(ii)** General conversions that are not within the metric system require the same procedure. The conversion factors used, however, do not simply move the decimal point (i.e., the exponent), but change both the coefficient and the exponent in the starting number.

Ex. 1. How many nickels are in $363.00?
Start with **dollars**; convert to **nickels.**

$\left(\frac{363\ \not{dollars}}{1}\right)\left(\frac{20\ nickels}{1\ \not{dollar}}\right) = (363)\ (20)\ nickels = 7{,}260\ nickels$

Ex. 2. The density of water is 1 g/cc. What is this equivalent to in terms of lb/in$^3$?

Solution:

This is a multistep conversion in which 2 units are to be converted.

g → lb.

cc. → in$^3$.

These problems are done by **completely converting one unit**, and then the other.

Recall 1 cc = 1 $cm^3$

$$\left(\frac{1.0\ \text{g}}{\text{cm}^3}\right)\left(\frac{1.0\ \text{lb}}{454\ \text{g}}\right)\left(\frac{2.54\ \text{cm}}{1\ \text{in}}\right)\left(\frac{2.54\ \text{cm}}{1\ \text{in}}\right)\left(\frac{2.54\ \text{cm}}{1\ \text{in}}\right) = \frac{(2.54)^3\ \text{lb}}{454\ \text{in}^3}$$

conversion from g to lb; conversion of $cm^3 \rightarrow in^3$

$$= 3.6 \times 10^{-2}\ \frac{\text{lb}}{\text{in}^3}$$

Ex. 3. Which of the following is the largest mass, and which is the smallest?

a) $3.5 \times 10^{-8}$ kg

b) 27 g

c) $4 \times 10^3$ mg

d) $3.4 \times 10^6$ μg

Solution:

The main difficulty in determining which quantity is the largest and which is the smallest occurs because each mass is expressed in a different unit. The first task then is to convert all the masses to the same unit. It is not important **which** units are finally compared (kg. with kg., g with g or other units). The question here will be answered by comparing the masses in units of **μgs.**

a. $$\left(\frac{3.5 \times 10^{-8}\ \text{kg}}{1}\right)\left(\frac{10^3\ \text{g}}{1\ \text{kg}}\right)\left(\frac{10^6\ \mu\text{g}}{1\ \text{g}}\right) = 3.5 \times 10^{-8+3+6}\ \mu\text{g}$$

$$= 3.5 \times 10^1\ \mu\text{g}$$

b. $$\left(\frac{27\ \text{g}}{1}\right)\left(\frac{10^6\ \mu\text{g}}{1\ \text{g}}\right) = 2.7 \times 10^7\ \mu\text{g}$$

c. $$\left(\frac{4 \times 10^3\ \text{mg}}{1}\right)\left(\frac{10^3\ \mu\text{g}}{1\ \text{mg}}\right) = 4 \times 10^6\ \mu\text{g}$$

d. $3.4 \times 10^6$ μg = $3.4 \times 10^6$ μg

The comparison is now between all masses expressed in μg.

a. $3.5 \times 10^1$ μg

b. $2.7 \times 10^7$ μg or $27 \times 10^6$ μg

c. $4 \times 10^6$ μg

d. $3.4 \times 10^6$ μg

$3.5\times10^{+1}\ \mu g < 3.4 \times 10^6\ \mu g < 4 \times 10^6\ \mu g < 2.7 \times 10^7\ \mu g$

increasing mass →

Therefore,

a. has the smallest mass (35 μg)

b. has the largest mass ($2.7 \times 10^7$ μg or 27,000,000 μg)

Work this problem again for practice and convert all the masses to a different common unit.

Exercise III: Perform the following conversions from the English to the metric system of units or vice versa. These conversions require the use of conversion factors and the factor label method. Refer to Table 1.3 (text) for the less widely used conversion factors.

| | | | | |
|---|---|---|---|---|
| 1. | 1.0 ft | ________ in | ________ cm | ________ mile |
| 2. | 300 g | ________ oz(avdp) | ________ lb | ________ μg |
| 3. | 6.3 fluid oz. | ________ μL | ________ L | ________ qt |

Ex.4. If the density of air is $1.25 \times 10^{-3}$ g/cc, what is the mass of the air in a room that is 5.0 meters long, 4.0 meters wide and 2.2 meters high?

Solution:

One approach to this problem is the following:

(i) The density is known in g/cc.

(ii) Determine the volume of the room in meters cubed ($m^3$) and then convert this volume to cc.

(iii) Knowing the density now in g/cc and the volume in cc, calculate the mass of the air in the room in grams.

(i) $D = 1.25 \times 10^{-3}$ g/cc

(ii) Determine the volume

Volume = length x width x depth = (5.0 m) (4.0 m) (2.2 m) = 44 $m^3$

Conversion of 44 $m^3$ to cc

$$\left(\frac{44\ \cancel{m^3}}{1}\right)\left(\frac{10^2\ cm}{1\ \cancel{m}}\right)\left(\frac{10^2\ cm}{1\ \cancel{m}}\right)\left(\frac{10^2\ cm}{1\ \cancel{m}}\right) = 44 \times 10^6\ cm^3 \text{ or } 4.4 \times 10^7\ cc$$

(iii) Determine the mass of air.

$$D = \left(\frac{M}{V}\right) \qquad M = (D)(V)$$

$$= (1.25 \times 10^{-3}\ g/cc)\ (4.4 \times 10^{7}\ cc)$$

$$M = 5.5 \times 10^{4}\ g$$

## THE CONCEPT OF SPECIFIC HEAT

The specific heat of a substance is an intrinsic property of a substance. It is a measure of how much energy (heat) is required to increase the temperature of one gram of a substance by 1°C.

Ex. 1 Consider the specific heat of 3 substances.

| | specific heat (cal/g-deg) |
|---|---|
| water | 1.00 |
| iron | 0.11 |
| lead | 0.038 |

a. Which substance requires the most heat to increase 1 g of the substance 1°C?

b. If equivalent amounts of heat (for ex., 100 cal) are added to 1 g of each substance, which substance will experience the largest increase in temperature? Which will experience the least increase?

c. If 60 cals. are added to 6 g of iron at 20°C, what will the final temperature be?

Solution:

a. Using the definition of specific heat directly, the values indicate that the specific heat for water requires 1.00 cal to increase 1 g of water by 1°C. The specific heat value for iron is about one tenth the value for water, while the value for lead is smaller yet. Therefore, by definition and simple comparison, water requires the most heat.

b. An equivalent amount of heat added to 1 g of each substance would increase the temperature of lead the most, while the increase in water would be the least. This is exactly the opposite order as observed in part a.

c. **Heat** absorbed by a substance will produce an increase in **temperature**. The magnitude of the increase will depend on the **specific heat** (S.H.) of the substance and the **mass**. As shown in the section 1.9 in the text, the equation relating these quantities is:

$$\text{heat} = (\text{S.H.}) \times (m) \times (T_2 - T_1)$$

This equation has 5 quantities (heat, S.H., m, $T_2$, $T_1$). In this problem and in others, you must first determine (i) which quantities are given and (ii) which quantity is to be determined (i.e., the unknown). Since only one quantity in this problem is to be determined, 4 quantities must be known.

Known quantities: heat = 60 cal

m = 6 g

S.H. (iron) = 0.11 cal/g-deg

$T_1$ = 20°C

$T_2$ = to be determined

Inserting these quantities into the equation

$$60 \text{ cal} = (0.11 \text{ cal/g-deg})\ (6\text{g})\ (T_2\text{-}20°)$$

Rearrange this equation by dividing both sides by (0.11 cal/g-deg) (6 g) and then canceling on the right side to yield

$$\frac{60 \text{ cal}}{(0.11 \text{ cal/g-deg})\ (6 \text{ g})} = (T_2\text{-}20°\text{C})$$

$$90° = T_2\text{-}20$$

Finally, add 20 to both sides to arrive at the final temperature, $T_2$.

$$110°\text{C} = T_2$$

Ex. 2. If 425 cal of heat are added to a sample of 52 g of copper at 25°C, what will be the final temperature?

$$\text{heat} = (\text{S.H.}) \times (\text{m}) \times (T_2\text{-}T_1)$$

Given:

heat = 1025 cal

m = 52 g

S.H. (copper) = 0.092 (Table 1.3, text)

$T_1$ = 25°C

$T_2$ = to be determined

$$425 \text{ cal} = 0.092 \text{ (cal/g-deg)} \times (52 \text{ g}) \times (T_2\text{-}25°\text{C})$$

$$\frac{425 \text{ cal.}}{(0.092 \text{ cal/g-deg})\ (52 \text{ g})} = T_2\text{-}25°\text{C}$$

$$89° = T_2\text{-}25°\text{C}$$

$$114° = T_2$$

## SELF-TEST QUESTIONS

MULTIPLE CHOICE. In the following questions, select the correct answer from the choices listed. In some cases two or more answers will be correct.

1. Which of the following are extensive properties of a substance?

   a. mass
   b. density
   c. length
   d. specific heat
   e. kinetic energy
   f. melting point

2. Indicate whether each statement is a fact, hypothesis or a theory.

   a. Smoking cigarettes is detrimental to your health.
   b. The ocean water at Miami, Florida contains more salt than the water in Galvaston, Texas.
   c. A micrometer (um) is larger than a nanometer (nm).

3. Which of the following are forms of matter?

   a. thought
   b. nucleus of an atom
   c. a vacuum
   d. a line drawn on a paper
   e. a shadow

4. The number $2.2 \times 10^{-4}$ is equivalent to:

   a. 0.0022
   b. $22 \times 10^{-5}$
   c. $22 \times 10^{-3}$
   d. $0.022 \times 10^{-2}$

5. The number $1 \times 10^{-4}$ is equivalent to:

   a. $\frac{1}{1 \times 10^{4}}$
   b. $10 \times 10^{-3}$
   c. 0.0001
   d. $100 \times 10^{-6}$

6. Which of the following numbers can be added directly in exponential notation?

   a. $3.26 \times 10^{-4}$
   b. $600 \times 10^{-2}$
   c. $32.6 \times 10^{-3}$
   d. $0.40 \times 10^{-4}$

7. Multiplication of $(3 \times 10^{-7})(1 \times 10^{6})(2 \times 10^{0})$ equals:

   a. $3 \times 10^{-1}$
   b. $6 \times 10^{-1}$
   c. 6
   d. $60 \times 10^{-3}$

8. The answer to the problem $\frac{93 \times 10^{4}}{3.0 \times 10^{-4}}$ is:

   a. $31 \times 10^{+8}$
   b. 31
   c. 13
   d. $31 \times 10^{0}$

9. The answer to the problem $\frac{(3 \times 10^{-7})(5 \times 10^{2})}{3 \times 10^{1} + 20}$ is:

a. $5 \times 10^{-4}$
b. $3 \times 10^{-6}$
c. $3 \times 10^{-4}$
d. 0.005

10. Which of the following are equivalent to 50 mL?

a. 0.5L
b. 50 cc
c. 0.05 L
d. $5 \times 10^{4}$ µL

11. Which of the following are equivalent to .5 ms?

a. 500 µs
b. $5 \times 10^{-4}$s
c. $8.3 \times 10^{-6}$ min
d. 5000 µs

12. A degree on the Centigrate scale is:

a. larger than the degree on the Fahrenheit scale
b. is the same size as the degree on the Kelvin scale
c. 9/5 larger than the Fahrenheit scale
d. none of the above

13. The temperature of -40°C corresponds to:

a. 313 K
b. 233 K
c. -40°F
d. -73°F

14. The temperature of 0 K corresponds to:

a. 273°C
b. absolute zero
c. -459.4°F
d. 32°F

15. A density of 1.5 g/cc is equivalent to:

a. 1500 g/L
b. 13.4 lb/ft$^3$
c. $3.3 \times 10^{-3}$ lb/mL
d. 6

16. How much energy is required to warm 15 g of ethyl alcohol from 4°C to 25°C? The specific heat of ethyl alcohol is 0.59 cal/g-deg.

a. 158 cal
b. 185 cal
c. 0.185 kcal
d. 37 cal

17. Which of the following ratios may be used directly as a conversion factor in converting microliters (μL) to liters (L)?

a. $\frac{1\text{ gal}}{3.785\text{ L}}$  c. $\frac{10^6 \mu\text{L}}{1\text{ L}}$

b. $\frac{1\text{ L}}{10^6 \mu\text{L}}$  d. $\frac{10^{-6} \mu\text{L}}{1\text{ L}}$

COMPLETION. Write the word, phrase or number in the blank which will complete the statement or answer the question.

1. Matter is defined as ______________________________.

2. Scientists proposed new ____________ to be tested and continually examine the validity of ____________ in more depth.

3. In the number, $4 \times 10^{-4}$, the number 4 is called the ____________, while -4 is the ____________.

4. Another way of writing the following numbers is:

a. $\frac{1}{10^{-4}}$ = ____________  c. $43 \times 10^6$ = ____________

b. 0.001 = ____________  d. 9,430,000,000 = ____________

5. The number $87{,}000 \times 10^{-16}$ is the same as $8.7 \times 10^{-n}$ in which n equals ____________.

6. SI is the abbreviation for ____________. This system is based on the ____________.

7. In the SI, the basic unit of length is the ____________, which is abbreviated as ________; the basic unit of volume is ____________, which is abbreviated as ____________.

8. A kilometer is ________ times longer than a centimeter.

9. Scientists use a balance to experimentally determine the __________ of a substance.

10. The basic unit for ____________ is the same in the English and metric systems. This abbreviation is ____________.

11. The size of the degree on the Kelvin scale is __________ to that on the Centigrade scale.

12. A temperature of 300°C corresponds to ________ K.

13. A temperature of 37°C is equal to ________ °F.

14. Absolute zero is defined as __________ K. At this temperature, the motion of molecules ____________.

15. There are __________ L in 6.7 gallons of wine.

16. The three states of matter are ________, ________ and ________.

17. To change a liquid to a gas, ____________ must be added to the liquid.

18. To change a liquid to a solid, energy must be ________ from the liquid.

19. For the three states of matter, the order of compressibility is ______ > ________ > ________.

20. Oil spilled by large tankers floats on the ocean because oil is _______ than ocean water.

21. The density and the specific gravity of a substance differ in that ____________.

22. Density and specific gravity are ____________ properties of a substance.

23. The density of a substance ____________ as the temperature is increased.

24. A hydrometer is used to measure the ___________ of a liquid.

25. Energy is defined as ____________

26. Kinetic energy is the energy of ____________, while potential energy is ____________ energy.

27. The law of conservation of energy states that ____________________________________

28. A high value for the specific heat of a substance indicates that ____________________.

29. Substance 1 has a specific heat of 3.0 cal/g-deg, while substance 2 has a specific heat of 1.0 cal/g-deg at 20°C. The addition of an equivalent amount of heat to 10 grams of either substance will result in substance ___________ increasing more in temperature.

30. In changing a large number into exponential notation form, the decimal point is moved to the _______ by n places and the exponent is equal to __________.

31. A temperature of 150 K is equal to ___________ on the Fahrenheit scale.

32. A volume of 131 mL of water weighs ____________.

33. A 15.0 mL volume of liquid with a specific gravity of 1.75 weighs ___________.

34. A volume of a clear blue liquid with a density of 0.89 g/cc weighs 43 g. The volume is ________. Its specific gravity is equal to _________.

35. A copper atom weighs 2.3 x $10^{-25}$ lb. Its weight in milligrams is _____________, and its corresponding weight in kilograms is ____________.

36. A drop or rain has a volume of about 50 microliters. This is equivalent to _________ quarts (1 quart = 0.946 L).

37. A single bond between two carbon atoms has a length of 1.54 Angstroms (1 Angstrom = 1 x $10^{-10}$ m). This bond length is equivalent to ___________ in.

38. A substance is found to have a density of 2.40 g/cc and weighs 100. g. Its volume is ___________.

39. At 20°C, the specific gravity of a solid substance A is 2.46, while the density of a liquid B is 2.0 g/cc. If solid A is placed in liquid B, it will __________.

40. A block of Blasa wood has dimensions of 2.00 cm x 6.00 cm x 1.50 m and a density of 0.120 g/cc. Its mass is __________.

41. Express the following quantities using the metric (SI) units indicated.

   a. 34 cm = __________ m

   b. $1 \times 10^3$ g = __________ μg

   c. $2.3 \times 10^3$ g/cc = __________ mg/L

   d. 6.3 cal/g-deg = __________ kcal/mg-deg

42. Volumes of liquids may be measured accurately with either a ________, ________ or ________. When only approximate volumes are needed, a __________ is used.

43. The ________ of an object is independent of location, while the ________ of an object is dependent on location.

44. In the laboratory, the ______ of a substance is measured on a balance (compare Fig. 1.4 and 1.5 in text).

45. Indicate the number of significant figures in the following

   a. 2.00 ______________

   b. 11,000 ______________

   c. 0.0043 ______________

   d. $5.32 \times 10^4$ ______________

   e. $5.0 \times 10^{-3}$ ______________

46. Carry out the following operations and report the correct number of significant figures in the answer.

   a.
$$\begin{array}{r} 14.00 \\ 1.042 \\ \underline{307.\phantom{000}} \end{array}$$

   b.
$$\begin{array}{r} 3{,}640.64 \\ -\underline{\phantom{0}13.1} \end{array}$$

   c. $0.37 \times 41.56 =$ ______________

   d. $\frac{18.4}{3.468} =$ ______________

e. $\frac{0.1340 \times 6.20}{146.00} =$ ______________________

## ANSWERS FOR FOCUSED REVIEW EXERCISES

Exercise I

1. $1.02 \times 10^{-2}$
2. $6.32 \times 10^{-12}$
3. $7.00 \times 10^{40}$
4. $1.3642 \times 10^{6}$
5. $1.1 \times 10^{24}$
6. $6 \times 10^{-21}$
7. $1.37 \times 10^{-8}$
8. $6 \times 10^{-3}$
9. $7.423 \times 10^{26}$
10. $4.361 \times 10^{30}$

Exercise II

1. $1.32 \times 10^{3}$ mL; $1.32 \times 10^{6}$ µL; $1.32 \times 10^{9}$ nL
2. $6.32 \times 10^{-3}$ g; $6.32 \times 10^{-6}$ kg; $6.32 \times 10^{-9}$ Mg
3. $3.9\times10^{-4}$ m; $3.9 \times 10^{-7}$ km; $3.9 \times 10^{2}$ µm
4. $3.6 \times 10^{-4}$ ms; $3.6 \times 10^{-7}$ s; $3.6 \times 10^{2}$ ns

Exercise III

1. 12 in; 30 cm; $1.9 \times 10^{-4}$ miles
2. 10 oz; 0.7 lb; $3 \times 10^{8}$ µg
3. $1.9 \times 10^{5}$ µL; $1.9 \times 10^{-1}$ L; $2.0 \times 10^{-1}$ qts

## ANSWERS TO SELF-TEST QUESTIONS

Multiple Choice

1. a,c,e
2. a (fact); b(hypothesis); c(fact)
3. b,d
4. b,d
5. a,c,d
6. a,d
7. b
8. a
9. b
10. b,c,d
11. a,b,c
12. a,b,c
13. b,c
14. b,c
15. a,c
16. b,c
17. b; in this conversion, L must be in the numerator

Completion

1. anything that has mass and also occupies space
2. hypotheses, theories
3. coefficient, exponent
4. a. $1 \times 10^{4}$; b. $1 \times 10^{-3}$; c. $4.3 \times 10^{7}$; d. $9.43 \times 10^{9}$
5. 12
6. System International, Metric System

7. meter, m, liter, L
8. $10^5$
9. mass
10. time, s
11. identical
12. 573 K
13. 98.6°F
14. 0 K or -273°C, ceases
15. 25 liters
16. solid, liquid, gas
17. heat energy
18. taken away
19. gas > liquid > solid
20. less dense
21. density is a physical quantity and has units, while specific gravity is a ratio of densities and is unitless
22. intensive
23. decreases
24. specific gravity
25. the capacity to do work
26. motion; stored
27. energy can neither be created or destroyed
28. the input of large amounts of energy or heat is necessary to increase the temperature of the substance
29. substance 2
30. left, plus n
31. -253°F
32. 131 g
33. 26.2 g
34. 48 cc (mL), 0.89
35. $1.0 \times 10^{-19}$ mg, $1.0 \times 10^{-25}$ kg
36. $5.3 \times 10^{-5}$ qt
37. $6.06 \times 10^{-9}$ in
38. 42 cc (mL)
39. sink
40. 216 g
41. a. 0.34 m; $1 \times 10^{+9}$ µg; $2.3 \times 10^9$ mg/L; $6.3 \times 10^{-6}$ cal/mg-deg
42. buret, pipet, volumetric flask, graduated cylinder
43. mass, weight
44. mass
45. a. 3
    b. at least 2; if all figures are significant, then 5
    c. 2
    d. 3
    e. 2
46. a. 322
    b. 3627.5
    c. 15
    d. 5.31
    e. 0.00569

Learning in old age is writing on sand but
learning in youth is engraving on stone.
- Arabian proverb

# 2

# Atoms

## CHAPTER OBJECTIVES

This chapter begins to present the fundamental aspects of chemistry. Whether or not you have been introduced to most of the terms and concepts previously in high school, you should very carefully master the material since it provides the real starting point and the basic grounding for subsequent chapters. You will find that an understanding of the biochemistry associated with a living organism relies heavily on the basic concepts and principles of chemistry.

After studying the chapter and working the assigned exercises in the text and the study guide, you should be able to do the following.

1. Summarize and contrast the ideas of Zeno, Democritus, Dalton and Bohr regarding the basic makeup of matter.
2. Name and write out the chemical symbols for the first twenty elements (this is a minimum objective which should be extended).
3. Distinguish between a pure compound and a mixture.
4. Describe four or five processes observed in everyday life which are examples of a physical or a chemical change.
5. State the four basic terms of Dalton's atomic theory.
6. Characterize the elementary particles in the atom, including their charge, mass and location within the atom.
7. Given the number of protons and neutrons in an atom, determine the charge of the nucleus and the atomic number and mass number for the element.
8. Given the atomic number of an element, locate and identify the element on the periodic table.
9. Distinguish between an atom, isotope, a positive and negative ion and a molecule.
10. Explain the difference between a representative element and a transition metal.

11. Write out the ground state electronic configuration for at least the first 40 elements, making use of Figure 2.13 (text), the Pauli Exclusion Principle and Hund's Rule.

12. Write out the ground state electronic configuration for positively and negatively charged ions of the first 20 elements.

13. Explain the trend in the ionization energies of the elements in a column (i.e., a group) and in a horizontal row on the periodic table.

14. List the four most abundant elements in (a) the earth's crust and those in (b) the human body.

15. Define the important terms and comparisons in this chapter and give specific examples where appropriate.

## IMPORTANT TERMS AND COMPARISONS

Atom, Ion and Molecule
Element
Compound and Mixture
Chemical and Physical Change
Chemical and Physical Property
Dalton's Atomic Theory
Law of Conservation of Mass
Law of Constant Composition
Diatomic Molecule
Proton, Neutron and Electron
AMU
Nucleons and Nucleus
Mass Number and Atomic Number
Atomic Weight and Relative Atomic Mass
Halogen
Group Numbers and Group Names
Representative Element and Transition Metal
Metal, Metalloid and Nonmetal
Noble Gases
Ground State and Excited State
s,p,d and f Orbitals
Shell and Subshell
Unpaired and Paired Electrons
Inner and Outer Shell
Electronic Configuration

## SELF-TEST QUESTIONS

MULTIPLE CHOICE. In the following exercises, select the correct answer from the choices listed. In some cases, two or more answers will be correct.

1. The concept of quantized energy levels in the atom was first proposed by:

   a. Zeno
   b. Bohr
   c. Lavoisier
   d. Democritus

2. Dalton's atomic theory includes the following statements:

   a. All atoms of the same element are identical to each other.
   b. All atoms are colorless.
   c. Atoms combine to form molecules. The molecule is a tightly bound group of atoms that acts as a unit.
   d. All matter is made up of very tiny indivisible particles called atoms.

3. The mass number of an atom which has 20 neutrons and a charge of +18 in the nucleus is:

   a. 20
   b. 38
   c. 56
   d. cannot determine

4. The atomic number of the atom in question 3 is:

a. 20
b. 38
c. 18
d. cannot determine

5. The ground state electronic configuration for the atom which has 6 protons and 6 neutrons in the nucleus is:

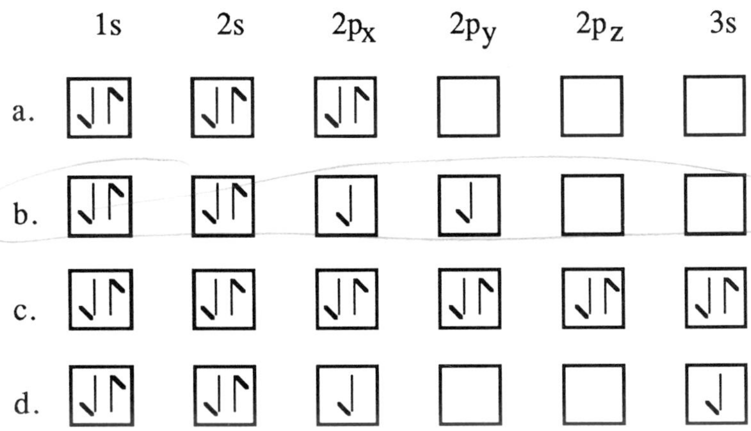


6. A possible excited state electronic configuration for the atom in question 5 is answer ________.

7. Which of the following elements is a metalloid?

a. Hydrogen, H
b. Germanium, Ge
c. Silver, Ag
d. Helium, He

8. Which of the following collection of elements are nonmetals?

a. Na, K, Rb
b. F, Cl, Br
c. O, S, Se
d. Cu, Ag, Au

9. The four most abundant elements in the earth's crust, listed in increasing order, are:

a. Ca < Fe < O < Si
b. Fe < Al < Si < O
c. N > C > H > O
d. Cl < O < N < P

10. The mass of a single atom of lead (Pb) is:

a. 3.5 g
b. 3.5 ng
c. $3.5 \times 10^{-22}$ g
d. $3.5 \times 10^{6}$ g

COMPLETION. Write the word, phase or number in the blank which will complete the statement or answer the question.

1. The three elementary particles in an atom are the ________, ________ and ________, with only ________ of them being classified as nucleons.

2. Color, melting point and density are examples of ________ properties of a compound.

3. The Law of Conservation of Mass states that ________________________________.

4. The atomic nucleus always possesses a __________ charge because the __________ resides in the nucleus.

5. Isotopes of an element have the same number of __________, but different number of __________. Therefore, their __________ numbers are the same although their __________ numbers differ.

6. A positive ion is produced from an atom by __________ an __________ the atom.

7. The halogens are in Group __________ and have __________ electrons in their outer shell.

8. Calcium, with atomic number 20, is in Group IIA. It is an example of a __________ element which has __________ electrons in its outer shell.

9. The ionization energy of elements in Group IA __________ going down the column, while the ionization energy going across a horizonal row of elements (from left to right) becomes __________.

10. The periodic table works because elements in the same __________ have the same __________ in the __________ shell.

11. In going down Group VIIA from flourine to astatine, the molecular weights progressively increase. Generally, as the molecular weights of similar compounds increase, the boiling points and melting points ________________.

## ANSWERS TO SELF-TEST QUESTIONS

### Multiple Choice

1. b
2. a, c, d
3. b
4. c
5. b
6. a or d
7. b
8. b, c
9. b
10. c

### Completion

1. proton, neutron, electron, 2
2. physical
3. matter cannot be created or destroyed
4. positive, proton
5. protons, neutrons, atomic, mass
6. pulling off or ionizing, electron from
7. VII, 7
8. representative or metallic, 2
9. decreases, greater
10. group or column, number of electrons, outer
11. increase

"I believe the chemical bond is not so simple as some people seem to think."
- Robert S. Mulliken

# 3

# Chemical Bonds

## CHAPTER OBJECTIVES

Although much remains to be learned about the atom, the titles of this and all subsequent chapters reveal that a great deal of attention is devoted to understanding the chemistry of ions and molecules. In either case, the outer **electronic configuration of the atom is altered** by the gain or loss of an electron to form a charged particle, or alternatively, electrons are shared in a common covalent bond between two atoms. As a result, the chemical and physical properties of these chemical species are different than the atoms from which they were derived. Also, in the case of complex ions or molecules, the species take on specific three-dimensional shapes, unlike that of the beautifully simple spherical shape of an atom. The molecular shapes can be predicted from the VSEPR theory.

After you have studied this chapter and worked the exercises in the text and study guide, you should be able to do the following.

1. Write the equations for the formation of a cation or anion from the parent atom.
2. Using the octet rule as a guide, predict why certain ions are stable, while others are nonexistent.
3. From a knowledge of the charge on the positive and negative ions and using the principle of electrical neutrality, predict the formulas for a variety of ionic compounds.
4. Distinguish between an atom, an ion and a molecule.
5. Describe the characteristic difference between an ionic bond and a covalent bond.
6. Characterize a single-, double- and triple-bond and give specific examples of compounds containing these bond types.
7. Write out the molecular formula for a variety of compounds, in addition to drawing the structural formula and Lewis structure for each compound.

8. Explain the difference between a covalent bond and a coordinate covalent bond.

9. State the VSEPR Theory. Given the number of valence shell electron pairs of the central atom of a molecule, predict the shape of the molecule and the size of the angle between the bonds in each. Give specific examples of molecules with each shape.

10. Define electronegativity and explain the trend in electronegativities in the periodic table as shown in Table 3.2 (text).

11. Draw the structural formulas for some di-, tri- and tetraatomic molecules. Given the electronegativity value for each atom, indicate whether the individual **bonds** and whether the **molecules** themselves are polar or nonpolar.

12. Using the general rules which relate the electronegativity differences of the atoms in a bond to the bond type, predict whether the bonds in a variety of diatomic molecules are ionic, polar-covalent or nonpolar-covalent.

13. Write out the name, together with the molecular and structural formulas for some common (Table 3.3, text) and some medicinally important (Box 3B, text) polyatomic ions.

14. Write out the names and chemical formulas for some of the inorganic compounds in the chapter.

15. Define the important terms and comparisons in this chapter and give specific examples where appropriate.

## IMPORTANT TERMS AND COMPARISONS

Ion, Cation, Anion
Ionic and Covalent Bonds
Electrical Neutrality
Molecular Orbital
Three-Dimensional Molecular Structure
Linear and Angular Molecules
VSEPR Theory
Shared and Unshared Electron Pair
Single, Double and Triple Bond

Molecular and Structural Formula
Coordinate Covalent and Covalent Bond
Coordination Compound
Coordination Number
Hemoglobin
Bond Angle
Trigonal, Pyramidal and Tetrahedral Shapes
Partial Charges, $\delta \pm$
Polar and Non-Polar Molecule

## SELF-TEST QUESTIONS

MULTIPLE CHOICE. In the following exercises, select the correct answer from the choices listed. In some cases, two or more answers will be correct.

1. The formula for cesium phosphate is:

   a. $Cs_3(PO_4)$
   b. $Cs_2(P_4)$
   c. $Cs(PO_4)_3$
   d. $CsPO_4$

2. The Lewis structure for the ozone molecule, $O_3$, is:

a. 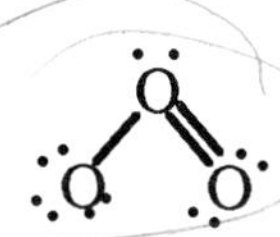

c. 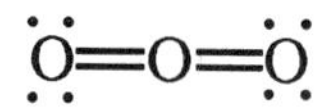

b. 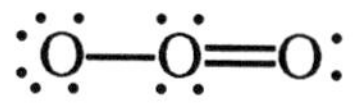

d. 

3. In which of the following can the two species react to form a coordinate covalent bond?

a. $H^+ + PCl_3$
b. two fluorine atoms
c. $BCl_3 + Cl^-$
d. $Ag^+ + H_2O$

4. Which of the following ions is stable?

a. $F^+$
b. $S^{2-}$
c. $Ca^+$
d. $O^{+2}$
e. $I^-$
f. $N^{3-}$

5. A particular atom has the following characteristics. The atomic number is 12 and it forms cations with a +2 charge. This element is:

a. Na
b. Al
c. S
d. Mg

6. Of the following bonds, the one with the greatest dipole moment is:

a. O-H
b. S-H
c. B-H
d. F-H

7. The order of decreasing dipole moments in the bonds in question 6 is:

a. H-F > O-H > S-H > B-H
b. H-F > S-H > B-H > O-H
c. O-H > H-F > S-H > B-H
d. S-H > O-H > H-F > B-H

8. Which of the following molecules have a correct Lewis structure?

a. 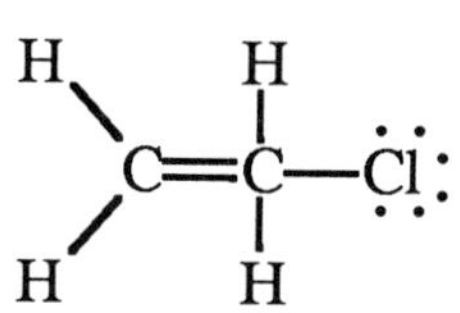

c. 

b. 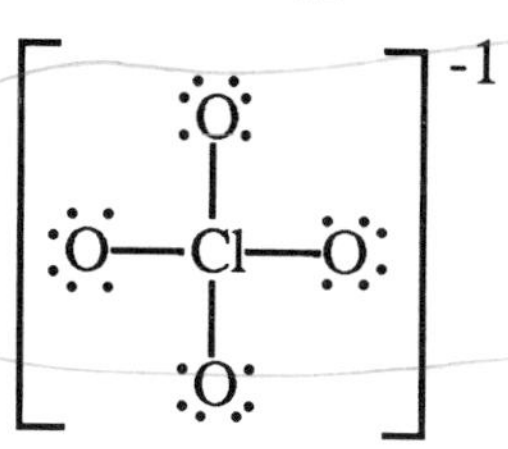

d. 

COMPLETION. Write the word, phrase or number in the blank which will complete the statement or answer the question.

1. Based on the VSEPR theory, determine the molecular shape of each molecule.

a. $AlF_4^-$ tetrahedral

b. $C_2HCl$ pyramidal

c. $H_3O^+$ linear

d. $PCl_3$ pyramidal

e. $BI_3$ trigonal

f. $I_3^-$ (The central I does not obey the octet rule.) linear

2. Classify whether the bond between the two elements shown is non-polar, covalent, polar covalent or ionic.

a. C, C nonpolar covalent

b. Li, F ionic

c. Na, O ionic

d. H, I nonpolar

e. O, S polar

3. Name these compounds or ions.

a. $CN^-$ cyanide anion

b. $(NH_4)_2SO_4$ ammonia sulfate

c. $NO_2^-$ nitrite ion

d. $K_2O$ potassium oxide

e. $MgCl_2$ magnessium chloride

f. $Li(CH_3COO)$ lithium acetate

4. Indicate the bond angles in these molecules.

a. $CO_2$ __________

b. $CCl_4$ __________

c. $H_2O$ __________

d. $BCl_3$ ______________________

e. $C_2H_2$ ______________________

5. Write out the chemical formulas for these compounds.

a. sodium bicarbonate Na HCO3

b. rubidium sulfate Rb2(SO4)

c. magnesium permanganate Mg(MnO4)2

d. ammonium oxide (NH4)2O

6. The formation of a positively charged cation from an atom requires the removal of an electron, while the formation of a negatively charged anion requires the addition of an electron.

7. A covalent bond involves the sharing of one or more electron pairs, while an ionic bond results from an electrostatic attraction between charged particles.

8. Although the $C^{+4}$ cation has a complete outer shell, it does not exist because concentrated charges on one atom are unstable.

9. In solid NaCl, each $Na^+$ has 6 $Cl^-$ ions as nearest neighbors ______________________.

10. An s orbital has a spherical shape, while a p orbital is shaped like a dumbells.

11. Name these compounds.

a. $NiCl_2$ Nickell (II) Chloride

b. $Cr(NO_3)_3$ Cromium (II) Nitrate

c. $NaH(SO_4)$ Sodium hydrogen sulfate

d. $MgF_2$ Magnessium Fluoride

12. Write the chemical formula for each of these compounds.

a. copper(II) iodide ______________________

b. sodium oxide ______________________

c. aluminum oxide ______________________

13. In the formation of the $F_2$ molecule, the $p_z$ atomic orbital from each of the two fluorine atoms overlap each other to form a molecular orbital . This molecular orbital, called a ____________ orbital, contains ____________ shared electrons which manifest the covalent bond in the molecule.

## ANSWERS TO SELF-TEST QUESTIONS

### Multiple Choice

1. a
2. a (note that 18 valence electrons are involved and the octet rule must be obeyed)
3. a, c, d
4. b, e, f
5. d
6. d
7. a
8. b, c, d

### Completion

1. a. tetrahedral
   b. linear
   c. pyramidal
   d. pyramidal
   e. trigonal planar
   f. linear

2. a. nonpolar covalent, $\Delta E = 0$
   b. ionic , $\Delta E = 3.0$
   c. ionic , $\Delta E = 2.4$
   d. nonpolar covalent, $\Delta E = 0.4$
   e. polar covalent , $\Delta E = 1.0$

3. a. cyanide anion (or ion)
   b. ammonium sulfate
   c. nitr_i_te anion (or ion)
   d. potassium oxide
   e. magnesium chloride
   f. lithium acetate

4. a. 180°
   b. 109.5°
   c. 105°
   d. 120°
   e. 180°

5. a. $Na(HCO_3)$
   b. $Rb_2(SO_4)$
   c. $Mg(MnO_4)_2$
   d. $(NH_4)_2O$

6. cation, removal, anion, addition
7. covalent, sharing, ionic, charged
8. concentrated charges on one atom
9. 6, nearest neighbors
10. s, dumbbell

11. a. nickel(II) chloride
    b. chromium(III) nitrate
    c. sodium hydrogen sulfate

d. magnesium fluoride

12. a. $CuI_2$
    b. $Na_2O$
    c. $Al_2O_3$

13. sigma, 2

....chemical transformations....are
responsible for corrosion and decay,
for development, growth and life.
- C.N. Hinshelwood
The Kinetics of Chemical Change, 1940

# 4

# Chemical Reactions

## CHAPTER OBJECTIVES

Chemistry is a quantitative science. This means that it is possible to determine the exact quantity of a substance - for example, the volume, density, mass or number of molecules of water in a raindrop. Importantly, one can relate **macroscopic** quantities, such as the mass of a substance, to **microscopic** quantities, such as the number of atoms or molecules in a given quantity of the substance. In addition, exact relationships can be represented by a balanced chemical equation or a mathematical equation. For example, we all realize that it requires gasoline to run an automobile. This process of combustion of the gasoline can be described quantitatively by a balanced chemical equation. This chapter presents these concepts, relationships and procedures which provide the basis for some of the quantitative aspects of chemistry.

After you have studied the chapter and worked the exercises in the text and study guide, you should be able to to do the following.

1. Calculate the formula weight of any ionic or molecular compound from its molecular formula.

2. Calculate the number of moles in a given mass of a compound.

3. Using Avogadro's number of $6.02 \times 10^{23}$, determine the number of molecules in a given mass of a compound.

4. Readily interconvert between the mass of a substance and the number of moles and molecules of the compound.

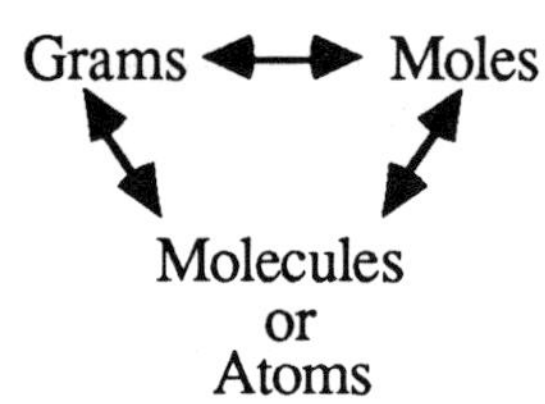


5. Using the atomic weights, determine the relative mass of pairs of atoms on the periodic table.

6. Balance simple chemical equations.

7. From a balanced chemical equation, carry out calculations to determine the mass and molar relationship between reactants and products.

8. Using the balanced equation and given the actual quantity of reactants and of the product formed in a reaction, calculate the percentage yield of product.

9. Write out the complete, balanced equations for the reaction of various ionic compounds in aqueous solution. Point out the spectator ions and then rewrite the ionic equation.

10. Be able to recognize and balance a redox reaction. Indicate the species which are oxidized and reduced and the species which are the oxidizing and the reducing agent, respectively.

11. Explain how a voltaic cell works.

12. Given balanced equations including the heat gain or loss, generally classify the reaction type and indicate whether the reaction is exothermic or endothermic.

13. Define the important terms and comparisons in this chapter and give specific examples where appropriate.

## IMPORTANT TERMS AND COMPARISONS

Chemical Reaction
Reactant and Product
Formula Weight and Molecular Weight
AMU
Avogadro's Number and Mole
Factor-Label Method
Conversion Factor
Balanced Equation
Coefficient and Subscript
Conservation of Mass
Actual Yield and Theoretical Yield
Side Reaction
Aqueous

Symbols, $\uparrow$, $\downarrow$
Oxidation and Reduction
Oxidizing and Reducing Agent
Combustion
Antiseptic and Disinfectant
Respiration
Rusting
Metabolism
Heat of Reaction and Heat of Combustion
Exothermic and Endothermic
Exergonic and Endogonic
Calorie

## FOCUSED REVIEW

To become comfortable and efficient with the mathematical manipulations of the new material, work the following exercises before going onto the SELF-TEST QUESTIONS.

Exercise I. Calculation of Formula Weights. From the molecular formulas, calculate the formula weight of each compound. Round the answer off to the first decimal place.

a. $Mg(ClO_4)_2$ ______________________

b. $Na_2O$ ______________________

c. $Al(H_3CCO_2)_3$ ____________________

d. $FeBr_3$ ____________________

e. $O_3$ ____________________

f. $C_{18}H_{36}O_2$ ____________________

g. $C_3H_7NO_2$ ____________________

h. $Pt(NH_3)_2Cl_2$ ____________________

<u>Exercise II.</u> Interconversion of Grams, Moles, and Molecules. The most important interrelationship in this chapter is that between grams, moles and molecules and is represented below.

Grams ↔ Moles

↕ ↕

Molecules or Atoms

In converting one of these quantities to another, a conversion unit is available and is used "as is" or inverted, depending on the direction of the conversion (i.e., whether converting from grams to moles or from moles to grams).

Using $H_2O$ as an example, it will be (or is) obvious that 1 gram of water is equivalent to the following.

$$1.0\ g\ H_2O \leftrightarrow 5.5x10^{-2}\ moles\ H_2O \leftrightarrow 3.3x10^{22}\ molecules\ H_2O$$

To accomplish these interconversions, the quantity is multiplied by the appropriate conversion factor, making use of the factor-label method. The conversion factor in each of these examples is underlined.

For example:

Given: 1.0 g $H_2O$ Convert to: moles $H_2O$

$$\left(\frac{1.0\ \cancel{g\ H_2O}}{1}\right)\underline{\left(\frac{1\ mole\ H_2O}{18\ \cancel{g\ H_2O}}\right)} = 5.5x10^{-2}\ moles$$

Given: 1.0 g $H_2O$ Convert to: molecules $H_2O$

$$\left(\frac{1.0\ \cancel{g\ H_2O}}{1}\right)\left(\frac{1\ \cancel{mole\ H_2O}}{18\ \cancel{g\ H_2O}}\right)\underline{\left(\frac{6.02x10^{23}\ molecules}{1\ \cancel{mole\ H_2O}}\right)} = 3.3x10^{22}\ molecules\ H_2O$$

Given: $5.5x10^{-2}$ moles $H_2O$ Convert to: g $H_2O$

$$\left(\frac{5.5x10^{-2}\ \cancel{moles\ H_2O}}{1}\right)\underline{\left(\frac{18\ g}{1\ \cancel{mole\ H_2O}}\right)} = 1.0\ g\ H_2O$$

Given: $5.5x10^{-2}$ moles $H_2O$ Convert to: molecules $H_2O$

$$\left(\frac{5.5x10^{-2} \text{ moles } H_2O}{1}\right)\left(\frac{6.02x10^{23} \text{ molecules}}{1 \text{ mole } H_2O}\right) = 3.3x10^{22} \text{ molecules } H_2O$$

Given: $3.3x10^{22}$ molecules $H_2O$ Convert to: moles $H_2O$

$$\left(\frac{3.3x10^{22} \text{ molecules } H_2O}{1}\right)\left(\frac{1 \text{ mole } H_2O}{6.02 \times 10^{23} \text{ molecules}}\right) = 5.5x10^{-2} \text{ moles } H_2O$$

Given: $3.3x10^{22}$ molecules $H_2O$ Convert to: g $H_2O$

$$\left(\frac{3.3x10^{22} \text{ molecules } H_2O}{1}\right)\left(\frac{1 \text{ mole } H_2O}{6.02x10^{23} \text{ molecules}}\right)\left(\frac{18 \text{ g } H_2O}{1 \text{ mole}}\right) = 1.0 \text{ g } H_2O$$

As noted in the text, the conversion from grams to moles or visa versa requires only one step, while conversion between grams and molecules requires 2 steps or 2 conversion factors. The importance of using the factor-label method is brought out in these calculations as well.

Carry out the following conversions, indicating the conversion unit used in the factor-label analysis. In addition, the answer should have the correct number of significant figures.

A. GRAMS ↔ MOLES CONVERSION

| | Grams | Conversion Factor | Moles |
|---|---|---|---|
| (i) | 4.0 g HF | ____________ | ____________ |
| (ii) | 33 g KOH | ____________ | ____________ |
| (iii) | 0.020 g $AlF_3$ | ____________ | ____________ |
| (iv) | ____________ | ____________ | 0.30 moles $PCl_3$ |
| (v) | ____________ | ____________ | 1.2 moles $GeH_4$ |
| (vi) | ____________ | ____________ | 0.01 moles $H_2CO_2$ |

B. MOLES ↔ MOLECULES CONVERSION

| | Grams | Conversion Factor | Molecules |
|---|---|---|---|
| (i) | 0.34 moles $H_2$ | ____________ | ____________ |
| (ii) | 11.3 moles $Na_2SO_4$ | ____________ | ____________ |
| (iii) | $3.4x10^{-24}$ moles $NH_3$ | ____________ | ____________ |
| (iv) | ____________ | ____________ | 1 molecule of $I_2$ |

| | | | |
|---|---|---|---|
| (v) | __________ | __________ | $6 \times 10^{20}$ molecules $O_3$ |
| (vi) | __________ | __________ | $4 \times 10^{6}$ molecules $HClO_2$ |

C. GRAMS ↔ MOLECULES CONVERSION

| | Grams | Conversion Factor | Molecules |
|---|---|---|---|
| (i) | 0.14 g $NCl_3$ | __________ | __________ |
| (ii) | 4.0 mg $S_8$ | __________ | __________ |
| (iii) | 7.4 ug HI | __________ | __________ |
| (iv) | __________ | __________ | $4 \times 10^{27}$ molecules $SeCl_2$ |
| (v) | __________ | __________ | $6 \times 10^{12}$ molecules $C_2H_4O_2$ |
| (vi) | __________ | __________ | $2 \times 10^{2}$ molecules $GeI_4$ |

Exercise III. Stoichiometry

In determining the weight relationships in chemical reactions, the key points are these:

1. The equation must be **balanced.**
2. **Molecules** react with **molecules**. **Moles** react with **moles**. The coefficients in the balanced equation dictate exactly how many of each of the reactants will react to form exactly how many product molecules. Therefore, the number of **moles** of the known reactant or product must be calculated first, if it is not given.
3. The coefficients in the balanced equation indicate the relative number of moles of each reactant and product. Therefore, multiply the moles of the known compound by the coefficient of the compound whose weight is unknown divided by the coefficient of the compound of known weight (or moles).
4. Convert the number of moles back to weight, by use of the appropriate conversion factor.

Consider the following equation. From the quantity given, determine the number of moles of the other quantities of interest.

$$C_7H_{16} + 11O_2 \rightarrow 7CO_2 + 8H_2O$$

A. Given: 0.1 mole $C_7H_{16}$

| moles $O_2$ | moles $CO_2$ | moles $H_2O$ |
|---|---|---|
| __________ | __________ | __________ |

B. Given: 0.1 mole $O_2$

| moles $C_7H_{16}$ | moles $CO_2$ | moles $H_2O$ |
|---|---|---|
| __________ | __________ | __________ |

C. Given 0.10 mole $CO_2$ produced

| moles $C_7H_{16}$ | moles $O_2$ | moles $H_2O$ |
|---|---|---|
| ______________ | ______________ | ______________ |

## SELF-TEST QUESTIONS

MULTIPLE CHOICE. In the following exercises, select the correct answer from the choices listed. In some cases, two or more answers will be correct.

1. The reaction of $C_4H_{10}$ with $O_2$ to produce $CO_2$ and $H_2O$ is what type of reaction?

   a. Dissociation
   b. Combustion
   c. Rusting
   d. Respiration

2. The formula weight for $Ca_3(PO_4)_2$ is:

   a. 279.3
   b. 310.3
   c. 167.3
   d. 543.3

3. A thimble of water contains $4.0 \times 10^{21}$ molecules. The mass of water is:

   a. 1.1 g
   b. 11 g
   c. $1.1 \times 10^{-8}$ g
   d. 0.11 g

4. Calculate the number of moles of aspirin, $C_9H_8O_4$, in a 4.0 g tablet.

   a. $2.3 \times 10^{-4}$
   b. 2.3
   c. $4.6 \times 10^{-3}$
   d. 0.023

5. The number of grams of $K_2O$ produced from the oxidation of 4.0 grams of K is:

   a. 4.7 g
   b. 4.0 g
   c. 9.4 g
   d. 6.3 g

6. If the reaction in problem 5 actually produces only 1.0 g $K_2O$, what is the percentage yield for the reaction?

   a. 25%
   b. 11%
   c. 21%
   d. 16%

7. In the equation below, the oxidizing agent in the reaction is:

   $$Al_{(s)} + H^+ = Al^{+3} + H_{2(g)}$$

   a. $Al_{(s)}$
   b. $H^+$
   c. $Al^{+3}$
   d. $H_2$

8. The species which is oxidized in problem 7 is:

   a. $Al_{(s)}$
   b. $H^+$
   c. $Al^{+3}$
   d. $H_2$

9. Indicate which reaction is the most endothermic.

   a. $A + B \rightarrow C + 1{,}600$ cal.
   b. $D + E + 12{,}000$ cal. $\rightarrow F$
   c. $G + H \rightarrow I + 2.3$ kcals.
   d. $X + Y + 2.6$ kcal. $\rightarrow Z$

10. A chemist used a laboratory calorimeter to determine that the combustion of glucose produced 686 Cal per mole of glucose. The amount of energy produced in the body by the metabolism of this same amount of glucose is:

   a. Greater than 686 Cal.
   b. Less than 686 Cal.
   c. Equal to 686 Cal.
   d. Can not predict

11. The correct answer to problem 10 derives from a law which states that the heat of reaction depends on the starting compounds and the final products and is independent of the reaction pathway. This law is called:

   a. Law of Balanced Reactions
   b. Avogadro's Law
   c. Hess's Law
   d. Hund's Rule

COMPLETION. Write the word, phrase or number in the blank which will complete the statement or answer the question.

1. The mass of germanium, Ge, (atomic number 32 and atomic weight 72.59) which contains a mole of Ge atoms is __________.

2. A __________ is used to determine the __________ value of foods.

3. The __________ ______ __________ is measured in a calorimeter. This is determined by measuring the temperature change of the water in the calorimeter. If the temperature decreases as a result of the reaction, the reaction is __________.

4. An __________ reaction is one in which __________ is given off, but not necessarily as __________.

5. __________ in organic compounds is defined as the gain of hydrogen or the loss of __________.

6. Four very common reaction types in aqueous solution which involve ions are (1) __________, (2) __________, (3) __________ or (4) __________.

7. Complete the following table.

| | Compound | Number of: Grams | Moles | Molecules |
|---|---|---|---|---|
| a. | $NH_3$ | __________ | $6.0 \times 10^{-4}$ | __________ |
| b. | $C_2H_2Br_2$ | 6.3 | __________ | __________ |

c. $Si_2C_6H_{18}$ __________ __________ $9.0x10^{12}$

8. Complete and balance the following equations.

a. $C_5H_{12} + O_2 \rightarrow CO_2 + H_2O$

b. $Al^{+3} + SO_4^{2-} \rightarrow$

c. $SnO_2 + H_2 = Sn_{(s)} + H_2O$

d. $Au + Cl_2 \rightarrow AuCl_3$

e. $C_6H_{12}N_4 + H_2O \rightarrow H_2CO + NH_3$

9. Consider the reaction of X and Y to form Z, with the indicated stoichiometry.

$$X + 2Y \rightarrow Z$$

The formula weights are: X = 50; Y = 200; Z = 100

a. The reaction indicates that __________ moles of X react with __________ moles of Y to produce __________ moles of Z.

b. If 200 grams of X is consumed in the reaction, the number of grams of Y consumed is __________ and the number of moles of Z produced is __________.

10. Consider the same stoichiometric equation in question 9. If 3 moles of Z were formed in the reaction, then __________ grams of X reacted with __________ grams of Y.

## ANSWERS FOR FOCUSED REVIEW EXERCISES

Exercise I.

a. 223.2
b. 62.0
c. 204.0
d. 295.5
e. 48.0
f. 284.0
g. 89.0
h. 300.0

Exercise II.

A. (i) $\frac{1 \text{ mole HF}}{20 \text{ g HF}}$ ; 0.20 moles

(ii) $\frac{1 \text{ mole KOH}}{56.1 \text{ g KOH}}$ ; 0.59 moles

(iii) $\frac{1 \text{ mole } AlF_3}{84 \text{ g } AlF_3}$ ; $2.4x10^{-4}$ moles

(iv) $\underline{137.4 \text{ g } PCl_3}$

$\frac{}{1 \text{ mole } PCl_3}$ ; 41 g $PCl_3$

(v) $\frac{76.6 \text{ g } GeH_4}{1 \text{ mole } GeH_4}$ ; 92 g $GeH_4$

(vi) $\frac{46 \text{ g } H_2CO_2}{1 \text{ mole } H_2CO_2}$ ; 0.46 g $H_2CO_2$

B. (i) $\frac{6.02 \times 10^{23} \text{ molecules}}{1 \text{ mole } H_2}$ ; $2.0 \times 10^{23}$ molecules

(ii) $\frac{6.02 \times 10^{23} \text{ molecules}}{1 \text{ mole } Na_2SO_4}$ ; $6.8 \times 10^{24}$ molecules

(iii) $\frac{6.02 \times 10^{23} \text{ molecules}}{1 \text{ mole } NH_3}$ ; 2.0 molecules

(iv) $\frac{1 \text{ mole } I_2}{6.02 \times 10^{23} \text{ molecules}}$ ; $4.2 \times 10^{-22}$ g.

(v) $\frac{1 \text{ mole } O_3}{6.02 \times 10^{23} \text{ molecules}}$ ; 0.48 g.

(vi) $\frac{1 \text{ mole } HC10_2}{6.02 \times 10^{23} \text{ molecules}}$ ; $4.6 \times 10^{-16}$ g.

C. (i) $\left(\frac{1 \text{ mole}}{120.4 \text{ g } NCl_3}\right)\left(\frac{6.02 \times 10^{23} \text{ molecules}}{1 \text{ mole}}\right)$ ; $7 \times 10^{20}$ molecules

(ii) $\left(\frac{1 \text{ mole}}{256.6 \text{ g } S_8}\right)\left(\frac{6.02 \times 10^{23} \text{ molecules}}{1 \text{ mole}}\right)$ ; $9 \times 10^{18}$ molecules

(iii) $\left(\frac{1 \text{ mole}}{127.9 \text{ g } HI}\right)\left(\frac{6.02 \times 10^{23} \text{ molecules}}{1 \text{ mole}}\right)$ ; $3 \times 10^{16}$ molecules

(iv) $\left(\frac{149.9 \text{ g } SeCl_2}{1 \text{ mole}}\right)\left(\frac{1 \text{ mole}}{6.02 \times 10^{23} \text{ molecules}}\right)$ ; $9.9 \times 10^{5}$ g

(v) $\left(\frac{60 \text{ g } C_2H_4O_2}{1 \text{ mole}}\right)\left(\frac{1 \text{ mole}}{6.02 \times 10^{23} \text{ molecules}}\right)$ ; $6 \times 10^{-10}$ g

(vi) $\left(\frac{580.02 \text{ g } GeI_4}{1 \text{ mole}}\right)\left(\frac{1 \text{ mole}}{6.02 \times 10^{23} \text{ molecules}}\right)$; $1.9 \times 10^{-19}$ g

Exercise III.

A. 1.1 moles $O_2$; 0.7 moles $CO_2$; 0.8 moles $H_2O$

B. $9 \times 10^{-3}$ moles $C_7H_{16}$; $6 \times 10^{-2}$ moles $CO_2$; $7 \times 10^{-2}$ moles $H_2O$

C. 0.01 moles $C_7H_{16}$; 0.16 moles $O_2$; 0.11 moles $H_2O$

# ANSWERS TO SELF-TEST QUESTIONS

## Multiple Choice

1. b
2. b
3. d
4. d
5. a
6. c
7. b
8. a
9. b
10. c
11. c

## Completion

1. 72.59 g
2. calorimeter, caloric value
3. heat of reaction, endothermic
4. exergonic, energy, heat
5. reduction, oxygen
6. precipitation, gas formation, acid neutralization of a base, redox

7. a. 0.010 g, $3.6 \times 10^{20}$ molecules

   b. $3.4 \times 10^{-2}$ moles, $2.0 \times 10^{22}$ molecules

   c. 2.2 ng, $1.5 \times 10^{-11}$ moles

8. a. $C_5H_{12} + 8O_2 \rightarrow 5CO_2 + 6H_2O$

   b. $2Al^{+3} + 3SO_4^{-2} \rightarrow Al_2(SO_4)_3$

   c. $SnO_2 + 2H_2 = Sn_{(s)} + 2H_2O$

   d. $Au + 3/2Cl_2 = AuCl_3$

   e. $C_6H_{12}N_4 + 6H_2O = 6H_2CO + 4NH_3$

9. a. one, two, one
   b. 1600 g, 4 moles

10. 150 g, 1200 g

Happy is he who gets to know the reasons for things.

- Motto of Churchill College, Cambridge

# 5

# Gases, Liquids and Solids

## CHAPTER OBJECTIVES

After studying the chapter and working the assigned exercises in the text and study guide, you should be able to do the following.

1. State the basic assumptions in the kinetic-molecular theory of gases. Indicate how each individual assumption helps in understanding the behavior or properties of ideal gases.
2. Explain how a barometer and a manometer are setup to read gas pressure.
3. Name the four properties of a gas which defines its physical state and indicate the common units for each.
4. Apply Boyle's Law ($P_1V_1 = P_2V_2$, conditions of constant temperature) to calculate the final volume or pressure of a gas.
5. Use Charles's Law ($V_1/T_1 = V_2/T_2$, conditions of constant pressure) to calculate the final volume or temperature of a gas.
6. Utilize Gay-Lussac's Law ($P_1/T_1 = P_2/T_2$, conditions of constant volume) to calculate the final temperature or pressure of a gas.
7. Apply the combined gas law ($P_1V_1/T_1 = P_2V_2/T_2$) to calculate the final pressure, temperature or volume of a gas.
8. Use the ideal gas law ($PV = nRT$) to calculate either the pressure, volume, the number of moles or the temperature of the gas.
9. Utilize the modified form of the ideal gas law to calculate the molecular weight or the mass of a quantity of gas.

10. Utilize Avogadro's Law (PV/RT = n, at conditions of a specific pressure and temperature) to show that equal volumes of gases at the same temperature and pressure contain equal numbers of gas molecules.

11. Apply Dalton's Law of Partial Pressures, ($P_T = P_1+P_2+P_3...$), to calculate the partial pressure of a gas in a mixture or the total pressure in a mixture of gases.

12. Use Graham's Law of Diffusion, $[(R_A^2/R_B^2) =[ M.W._B / M.W._A]$, to calculate either the relative rates of diffusion of two gases or the molecular weight of one of the gases.

13. Name, define and give the approximate energies involved in the three basic intermolecular forces between molecules.

14. Draw a few molecules which can participate in intermolecular hydrogen bonding and show the specific interactions involved.

15. Define boiling point and explain the factors which influence the temperature at which a liquid boils.

16. Distinguish between amorphous and crystalline solids.

17. Draw a heating curve for ice, explain the main features of the curve and explain the significance of the heat of fusion and the heat of vaporization in this figure.

18. Define the important terms and comparisons in this chapter and give specific examples where appropriate.

Make a summary sheet of the important laws and corresponding equations presented in this chapter which provide the basis for characterizing the property of a gas. Indicate the parameters which are kept constant in each relationship and the units of each.

## IMPORTANT TERMS AND COMPARISONS

Gaseous, Liquid and Solid State
Melting Point and Freezing Point
Barometer and Manometer
Torr, mm Hg, Pascal
Atmosphere (unit of pressure)
Systolic and Diastolic Pressure
Sphygmomanometer
Boyle's, Charles's and Gay-Lussac's Laws
22.4 L
Partial and Total Pressure
Dalton's and Graham's Laws
Arterial and Venous Blood
London Dispersion Force and Hydrogen Bond
Fluidity
Vapor Pressure of a Liquid
Boiling Point and Melting Point
Crystallization
Crystals, Amorphous and Network Solids
Phase Change
Heating Curve
Heat of Fusion and Heat of Vaporization

## FOCUSED REVIEW

One of the most perplexing difficulties that many students encounter in problem solving is deciding the equation to use to obtain the answer. Since there are a number of equations (relationships) which may be used in solving gas law problems, a simple procedure is necessary to find just the right relationship for the particular problem.

The parameters that may be involved in a particular problem are the following:

| Parameter | Common Units |
|---|---|
| Pressure (P) | Atmosphere, mm Hg, Torr |
| Volume (V) | Liter, milliliter |
| Temperature (T) | °C, K |
| Moles (n) | -------- |
| Mass (m) | Grams |
| Molecular Weight (M.W.) | Atomic Mass Units |

The types of problems can be subdivided into two major categories. The first is the situation in which the gas is completely characterized in one state, the **conditions are CHANGED**, and one of the characteristics of the second state is unknown. A number of situations may arise.

TABLE I

| | State 1 Known Quantity | | State 2 Known Quantity | State 2 Unknown Quantity |
|---|---|---|---|---|
| Ideal Gas Law: | $\frac{P_1V_1}{T_1} = \frac{P_2V_2}{T_2}$ | | | |
| | $(P_1,V_1T_1)$ | ⟶ | 1. $P_2V_2$ | $T_2$ or |
| | | | 2. $P_2,T_2$ | $V_2$ or |
| | | | 3. $V_2,T_2$ | $P_2$ |
| Boyle's Law: | $P_1V_1 = P_2V_2$, at constant T | | | |
| | $P_1,V_1$ | ⟶ | 1. $P_2$ | $V_2$ or |
| | | | 2. $V_2$ | $P_2$ |
| Charles's Law: | $\frac{V_1}{T_1} = \frac{V_2}{T_2}$, at constant P | | | |
| | $(V_1,T_1)$ | ⟶ | 1. $V_2$ | $T_2$ or |
| | | | 2. $T_2$ | $V_2$ |
| Gay-Lussac's Law: | $\frac{P_1}{T_1} = \frac{P_2}{T_2}$, at constant V | | | |
| | $(P_1,T_1)$ | | 1. $P_2$ | $T_2$ or |
| | | | 2. $T_2$ | $P_2$ |

The second category of problems involves a gas or gases in one **PARTICULAR** state. **No changes in conditions are involved.** In this case, a specific property in this state is to be calculated, using either the ideal gas law, Dalton's Law of partial pressures, Avogadro's Law or Graham's Law of diffusion.

Recall that the ideal gas law equation may be written in two forms.

$$PV = nRT \qquad (1)$$

and since

$$n = \frac{g}{M.W.}$$

$$PV = \frac{gRT}{(M.W.)} \qquad (2)$$

Therefore, one can solve for any one of the parameters in equation (1) or (2), if the other quantities are given in the problem.

Two final points of interest.

1. R, the universal gas constant is a physical quantity, which you'll recall has both a numerical value and **units**. The value of R presented in the text is R = 0.082 L-atm/mole-K. Therefore, it is essential in using this equation with this value of R that the gas parameters conform with these units of R. That is, the parameters must have units of P (atm), V (L), n (moles) and T (K). If the parameters are not given in the proper units, convert them immediately. Be on the outlook for just this situation.

2. In problems involving two states of a gas and using any one of the four gas laws presented, it is also essential that the units characterizing state 1 be the same as in state 2. For example, if the pressure in state 1 is given as 1.2 atm, while the pressure in state 2 is given as 500 mm Hg, one of these values (usually the 500 mm Hg value) must be converted to the other before the calculation.

Knowing these preliminaries and precautions, the procedure for analyzing and solving a gas law problem is:

1. Determine if the problem involves the gas in only one state or two different states.
2. Determine (**write out**) what is known and what is to be determined (unknown) in the problem.
3. From your analysis in steps 1 and 2, decide which law and equation is appropriate to solve the problem.
4. Rearrange the equation if necessary and check that all units are appropriate.
5. Solve for the unknown quantity.

To pull some of this together, consider the following problem.

If a sample of 1.78 moles of $CHCl_3$ gas occupies 68 L at a pressure of 275 mm Hg, what is its temperature in °C?

Step 1. The problem involves $CHCl_3$ gas in **one state**. There are no changes to a new state. Therefore the equations as written in Table I (Study Guide) are not appropriate and either the (1) ideal gas law, (2) Dalton's Law of partial pressures, (3) Avogadro' Law or (4) Graham's Law of diffusion will be used.

Step 2. Knowns / Unknown

| Knowns | Unknown |
|---|---|
| n = 1.78 moles | |
| V = 68 L | T = ? |
| P = 275 mm Hg | |

Step 3. From the parameters given and the unknown quantity, the ideal gas law should be used in the form.

$$PV = nRT$$

Step 4. Rearrange by dividing through each side by nR. This yields

$$T = \frac{PV}{nR}$$

Convert pressure (P = 275 mm Hg) into units of atmosphere to conform with units of R, the universal gas constant.

$$P = \left(\frac{275 \text{ mm Hg}}{1}\right)\left(\frac{1 \text{ atm}}{760 \text{ mm Hg}}\right) = 0.362 \text{ atm}$$

Step 5. $T = \dfrac{(0.362 \text{ atm})(68 \text{ L})}{(1.78 \text{ moles})(0.082 \text{ L-atm/mole-K})} = 169 \text{ K}$

The final step in this problem is to convert the temperature from K to °C.

°C = 273° + K = 273° + 169,

Therefore, the final answer is:

**T = 442°C**

SELF-TEST QUESTIONS

MULTIPLE CHOICE. Select the correct answer from the choices given. Only one of the choices is correct.

1. A closed aerosol-can which had been filled with shaving cream contains 3.0 g of nitrogen gas ($N_2$) as the propellant. If the volume of the can as 200 mL, the pressure in the can at 25°C is:

   a. 1.1 atm
   b. 130 atm
   c. 1.3 atm
   d. 13 atm

2. If the can in question 1 is thrown into an incinerator or fire, the temperature in the can could well get as high as 850 K. The pressure in the can, assuming that it did not explode, would be:

   a. 37 atm
   b. 3.7 atm
   c. 370 atm
   d. 370 torr

3. A gas, which is initially contained in a movable cylinder of 900 mL volume and at 25°C., exerts a pressure of 3 atmospheres. The pressure is reduced at constant temperature to a value of 380 mm Hg. The new volume is:

| | | | |
|---|---|---|---|
| a. | 5400 L | c. | $7.1 \times 10^{-3}$ L |
| b. | 5.4 L | d. | 7.1 L |

4. A piece of Dry Ice (solid carbon dioxide, $CO_2$) is placed in a 1 L tank. It sublimes to $CO_2$ gas at STP. The amount of $CO_2$ was:

| | | | |
|---|---|---|---|
| a. | 1 mg | c. | 4.4 g |
| b. | 44 g | d. | 2 g |

5. The average kinetic energy of molecules is proportional to the

| | | | |
|---|---|---|---|
| a. | temperature in °C | c. | temperature in K |
| b. | velocity of the molecule | d. | volume at STP |

6. The official S1 unit of pressure is:

| | | | |
|---|---|---|---|
| a. | torr | c. | atmosphere |
| b. | pascal | d. | mm Hg |

7. The autoclave operates on the principle of:

   a. Dalton's Law of Partial Pressure
   b. Boyle's Law
   c. Gay-Lussac's Law
   d. Charles's Law

8. A gas at 25°C is initially contained in a movable cylinder of 2000 mL volume and exerts a pressure of 38 mm Hg. The pressure is changed to 0.5 atm and the volume of the cylinder increases to 2.2 L. The final temperature in the cylinder is:

| | | | |
|---|---|---|---|
| a. | $3 \times 10^3$ K | c. | 3.2 K |
| b. | 275 K | d. | $2 \times 10^4$ K |

9. Consider the reaction: $3K + Y \rightarrow 2Z$. At STP, 2 L of Z are produced from the reaction of X and Y. The number of liters of X necessary is:

| | | | |
|---|---|---|---|
| a. | 1 L | c. | 3 L |
| b. | 2 L | d. | cannot determine |

10. A quantity of 6 grams of a gas occupies 300 milliliters at STP. The molecular weight of the gas is:

| | | | |
|---|---|---|---|
| a. | 440 | c. | 44 |
| b. | 660 | d. | 66 |

<u>COMPLETION</u>. Write the word, phrase or number in the blank which will complete the statement or answer the question.

1. STP indicates conditions in which T = __________ and P = __________.

2. The boiling point of HF (M.W. = 18) is higher than that for $CH_4$ (M.W. = 18) because of __________, while the boiling point of n-octane is higher than that for propane because of the __________.

3. The behavior of an ideal gas at constant volume but at varying pressure or temperature is governed by __________ Law.

4. The $O_2$ gas in a 1 L tank at 1 atmosphere is transferred to a 2 L tank which already contains 2 atmospheres of $N_2$ gas. The final pressure is __________.

5. The higher the temperature, the __________ the __________ energy of molecules. Therefore, by heating a closed container of gas, the pressure will __________.

6. A 5.6 L container of Ne or $O_2$ contains __________ moles of gas or __________ molecules at STP.

7. The energy required to break a hydrogen bond is approximately __________ kcal/mole.

8. A molecule used in perfume is determined to diffuse one-tenth as fast as neon gas. Its molecule weight is __________.

9. The basic postulates in the kinetic molecular theory of gases are:

   i. ______________________________

   ii. ______________________________

   iii. ______________________________

   iv. ______________________________

   v. ______________________________

   vi. ______________________________

10. The melting point of diamond, which is a __________, is much higher than that for a __________ crystal.

11. The two plateaus in a heating curve represent the temperatures at which two phases __________. The amount of heat necessary to convert one gram of water at 100°C to steam is called the __________.

12. The intermolecular forces responsible for the solidification of Xe and other noble gases are __________.

13. The unit of pressure, __________, is equivalent to the __________.

14. Convert the physical quantities in the left column to their equivalent in the right-hand column.

    a. 4.3 mL ______________________ L

    b. 34°C ______________________ K

    c. 450 mm Hg ______________________ atm

d. 43 g $O_2$ __________ moles $O_2$

e. 372 K __________ °C

f. STP __________ torr

__________ °C

g. 17 cal/mole __________ Kcal/mole

15. Match up the unit or instrument in the column at the right with its use in the left-hand column.

| Unit or Instrument | Use |
|---|---|
| a. Barometer | i. Measure blood pressure |
| b. Sphygmomanometer | ii. Laboratory instrument to measure pressure |
| c. Closed container of liquid | iii. Hospital unit used for patients with CO poisoning, smoke inhalation, in addition to other uses |
| d. Hyperbaric Chamber | iv. Measures atmospheric pressure |
| e. Manometer | v. Used in determination of vapor pressure |

16. Water has a superstructure involving intermolecular __________, which in part explains why the density of __________ is greater than that of __________.

17. As the temperature decreases, the vapor pressure of a liquid __________.

18. __________ is the transition from the solid state directly into the __________ state without going through the __________ state.

19. An amorphous solid differs from a crystalline solid in that the molecules or ions are solidified in a __________ pattern.

20. The ideal gas equation is written __________.

## ANSWERS TO SELF-TEST QUESTIONS

Multiple Choice

1. d
2. a
3. b
4. d
5. c
6. b
7. c
8. a
9. c
10. a

Completion

1. T = 0°C, P = 1 atm
2. intermolecular hydrogen bonds, (higher) molecular weight
3. Gay-Lussac's Law
4. 2.5 atm
5. greater, kinetic, increase
6. 0.25, $1.5 \times 10^{23}$
7. 5-10

8. 2000
9. These statements can be paraphrased.
   i. A gas consists of molecules traveling in random directions, in straight lines and at a range of speeds.
   ii. The average kinetic energy, and therefore the average velocity of gas molecules, is directly proportional to the absolute temperature.
   iii. Gas molecules collide with each other and may exchange energy between each other, but the total kinetic energy is conserved.
   iv. The molecules in a gas take up essentially no volume. The volume of the gas is the volume of the container it is in.
   v. There are no attractions between the gas molecules.
   vi. The pressure of the gas is due to collisions of the molecules with the walls of the container. The greater the number of collisions per unit time, the greater the pressure.
10. network solid, molecular
11. coexist, heat of vaporization
12. London dispersion forces
13. torr, mm Hg or also 1 atm = 101,325 pascals is acceptable. The former equivalence is more commonly used, however.
14. a. $4.3x10^{-3}$ L
    b. 307 K
    c. 0.592 atm
    d. 1.3 moles $O_2$
    e. -1°C
    f. 760 torr, O°C
    g. $1.7x10^{-2}$ Kcal/mole
15. a. iv
    b. i
    c. v
    d. iii
    e. ii
16. hydrogen bonding, liquid water, ice
17. decreases
18. sublimation, gaseous, liquid
19. random
20. PV = nRT

The fluids of the human body and the seas of the earth contain common table salt, sodium chloride, and an assortment of other chemicals.

- S. Brooks
The Seas Inside Us: Water in the Life Processes

# 6

# Solutions and Colloids

## CHAPTER OBJECTIVES

After studying the chapter and working the assigned exercises in the text and study guide, you should be able to do the following.

1. Summarize the characteristics of a mixture, a suspension, a colloidal dispersion and a true solution and give specific examples of each.

2. Indicate the factors which influence the solubility of a (1) gas, (2) liquid or (3) solid in a liquid solvent.

3. Perform calculations to determine the concentration of a solution expressed in (1) molarity, (2) % w/v and (3) w/w. Given the concentration of a solution in any of these forms, convert to its equivalent in the other two.

4. Using the formula, moles = $M_1 \times V_1 = M_2 \times V_2$, calculate the concentration of solutions made by the dilution of a more concentrated solution.

5. Outline the characteristics of water which are most responsible for the unusual properties, such as (1) its high boiling point and (2) its ability to dissolve ionic compounds.

6. Specify the fundamental difference between a strong and weak electrolyte and a non-electroyte.

7. Discuss the mechanisms by which water can effectively dissolve particular covalent molecules.

8. Considering both molecular and ionic compounds, carry out calculations to determine the freeezing point depression produced by the various aqueous solutions.

9. Define osmosis and osmolarity and discuss how these concepts relate to specific areas of medicine.

10. Define the important terms and comparisons in this chapter and give specific examples where appropriate.

## IMPORTANT TERMS AND COMPARISONS

Homogeneous and Heterogeneous Mixtures
Solute, Solvent and Solution
Clear vs. Colorless
A Physical Constant
Saturated, Unsaturated and Supersaturated Solutions
"Like Dissolves Like"
Thermal Pollution
Concentration
Dilute vs. Concentrated
Percent Concentration
(% w/v), (w/w) and (v/v)
Molarity and Osmolarity
Proof
Solvation
Plaster of Paris
Electrolyte and Non-Electrolyte
Strong and Weak Electrolyte
Solution and Suspension
Transparent, Translucent and Opaque
Solvation Layer
Physiological Saline
Isotonic, Hypertonic and Hypotonic
Hemolysis and Crenation
Dialysis and Hemodialysis

## FOCUSED REVIEW OF CONCEPTS

In all chemically related fields, we are continuously called upon to make observations or measurements and to characterize a mixture or solution, either in a qualitative or quantitative sense. Recall that chemistry is an experimental science and our current knowledge is derived from **careful observations** of experimental happenings. Concepts in this chapter lay the foundation for such characterization. Consider, for example, that you are presented with a test tube containing some liquid and asked to characterize it as completely as you can. Your preliminary, yet very important, judgments will necessarily be **qualitative**. Initially, you might focus on these questions.

1. Does the vial contain a true solution, a colloidal dispersion or a suspension?

2. If the vial contains a true solution,

   a. Is the solution clear (transparent), translucent or opaque?
   b. Is the solution colored or colorless?
   c. Does the solution conduct electricity?
   d. Is the liquid simply a solvent or an actual solution?
   e. If it is a solution, is it dilute or concentrated and is it unsaturated, saturated or supersatured with solute?

At this point, it should be evident that although the characteristics are qualitative, we know a great deal about the unknown solution in the test tube.

To characterize the solution further, however, requires **quantitation**. The following table collects a number of means of quantitating solutions, with the definition and examples of each. However, first just focus your attention on a basic point which is often times not clearly defined and leads to difficulty. You will note that all the quantitative concentration terms in Table I are expressed as the quantity of one reagent in a certain volume of **SOLUTION** (= solute plus solvent). The emphasis here is to distinguish the difference between the volume of the total solution from the volume of the solvent used to dissolve the solute.

TABLE I

Quantitative Description of Solutions

| Term | Definition | Examples |
|---|---|---|
| Molarity (M) | Moles solute per 1 L **SOLUTION** M = (moles/L) | 3 M NaCl; Dissolve 175 g NaCl in enough water to make 1 L solution. |
| Osmolarity | Molarity times the number of particles (n) produced by each mole of solute<br><br>Osmolarity = n x M | 0.1 M NaCl; Osmolarity = 0.2 $NaCl \rightarrow Na^+ + Cl^-$; n = 2<br>0.3 M $Mg_3(PO_4)_2$; Osmolarity = 1.5 $Mg_3(PO_4)_2 \rightarrow 3Mg^{+2} + 2PO_4^{3-}$ (n = 5)<br>0.8 M glucose (n = 1) Osmolarity = 0.8 |
| (% w/v) | Weight solute per 100 mL **SOLUTION** | 5 g of X in 100 mL solution 5 % w/v |
| (% w/w) | Weight solute per weight **SOLUTION** | 5 g of X in 50 g solution 10 % w/w |
| (% v/v) | In a 2 component solution, volume of A added to enough liquid B to make 100 mL **SOLUTION** | 17 mL liquid A plus enough liquid B to make 100 mL solution 17 % v/v |

SELF-TEST QUESTIONS

MULTIPLE CHOICE. In the following exercises, select the correct answer from the choices listed. In some cases, two or more choices will be correct.

1. Which of the following gases can dissolve in rain droplets to produce acid rain.

   a. $O_2$
   b. $SO_3$
   c. $CO_2$
   d. CO

2. The solubility of a solid solute in water will depend on what factors?

   a. temperature
   b. size of the solute particles
   c. polarity of solute
   d. pressure

3. Which of the following covalent molecules can hydrogen-bond to water?

   a. nitrogen, $N_2$
   b. ammonia, $NH_3$
   c. methane, $CH_4$
   d. ethyl alcohol, $C_2H_5OH$

4. Indicate which of the following properties a colloidal dispersion has that a true solution does not.

a. homogeneity
b. Tyndall effect
c. filterable with ordinary paper
d. electrical conductance

5. An example of a colloid in which a solid is dispersed in a liquid is:

a. cheese
b. whipped cream
c. jellies
d. clouds

6. Which of the following properties does a suspension **not** have?

a. settles on standing
b. homogeneous
c. translucent or opaque
d. particles size of < 1000 nm

7. A solution containing proteins and NaCl salt is placed in a semipermeable membrane bag and dialyzed against distilled water. Which component or components will be found in the distilled water?

a. NaCl, proteins
b. NaCl
c. proteins
d. none of the components

8. The solubility of a polar gas molecule in a liquid solvent increases with:

a. increasing temperature
b. decreasing temperature
c. increased gas pressure
d. increased solvent polarity

9. The preparation of 300 mL of a $2.0 \times 10^{-5}$ M solution of hemoglobin (molecular weight 64,000) requires how many grams of hemoglobin?

a. 0.38 g
b. 38 g
c. $6.4 \times 10^{-1}$ g
d. 7.6 mg

10. The reaction of oxygen gas with hemoglobin (Hb) is shown in the equation below:

$$Hb + 4O_2 = Hb(O_2)_4$$

How many moles of $O_2$ can react with the hemoglobin solution in problem 9?

a. $3 \times 10^{-2}$ moles
b. $2.4 \times 10^{-5}$ moles
c. $6 \times 10^{-6}$ moles
d. $3 \times 10^{-6}$ moles

11. A solution of HCl is prepared to be $1 \times 10^{-2}$ M. A 15 mL aliquot is taken and diluted with 80 mL of water. The final concentration of HCl is

a. $1.9 \times 10^{-4}$ M
b. $1.9 \times 10^{-3}$ M
c. $1.6 \times 10^{-3}$ M
d. $1.6 \times 10^{-4}$ M

12. A 1.5 M solution is made by dissolving 50 g of compound X in 50 mL of solution. The molecular weight of X is :

a. 660
b. 100
c. 150
d. 300

COMPLETION. Write the word, phrase or number in the blank space or draw the appropriate structure in answering the question.

1. Preparation of 500 mL of a 0.5 M NaCl solution requires __________ NaCl.

2. The osmolarity of a 0.1 M $Na_3(PO_4)$ solution is __________. The freezing point of this solution is determined on a 50 ml portion (aliquot) of the solution. The observed freezing point will be __________.

3. Crystalline substances which contain no water are called __________. Substances which become hydrated on exposure to the open air are referred to as __________ compounds.

4. Three 0.1 M solutions are prepared, one with HF, another with NaCl and the last one with sucrose (common sugar). Because the labels came off the bottles, we shall identify the solutions as simply solutions A, B and C. Solutions marked A and B conduct electricity; however, solution A clearly has a much greater capacity to conduct electricity than does solution B. Solution C does not conduct electricity. Identify each solution and characterize it according to its electrolytic character.

| Compound | Solution | Type of Electrolyte |
|---|---|---|
| HF | (i) | (iv) |
| NaCl | (ii) | (v) |
| Sucrose | (iii) | (vi) |

5. A true solution contains particles with a maximum diameter of __________.

6. A __________ solution has a greater osmolarity than do red blood cells. Therefore, placing red blood cells in this solution will cause the cells to __________. A 0.3 M glucose solution is a __________(tonic) solution.

7. The concentration of a 0.40 M glucose, $C_6H_{12}O_6$, solution, can be expressed in terms of (% w/v). The equivalent concentration expressed in this way would be __________ (% w/v).

8. The random motion of a colloid suspended in a solvent is called __________.

9. A biochemist prepares a solution containing hemoglobin, the compound in the red blood cells which carries $O_2$ to our cells. The hemoglobin in this solution is the __________.

10. A student adds benzene to a bottle containing $CCl_4$ and also to a separate bottle containing methyl alcohol, $CH_3OH$. The benzene is miscible in the __________, but immiscible in the __________. This is an example of the rule that "like dissolves like." In this case, a __________ compound dissolves in a __________ solvent.

11. Osmosis involves the passsage of __________ from the __________ to the __________ side of a semipermeable membrane.

12. The __________, or kidney cells act as the filtering unit in our bodies.

## ANSWERS TO SELF TEST QUESTIONS

### Multiple Choice

1. b, c
2. a, c
3. b, d
4. b
5. c
6. b, d
7. b
8. b, c, d
9. a
10. b
11. c
12. a

### Completion

1. 14.6 g
2. 0.4; -0.76°C
3. anhydrous; hygroscopic
4. (i) Solution B (ii) Solution A (iii) Solution C (iv) weak electrolyte (v) strong electrolyte (vi) non-electrolyte
5. 1 nm
6. hypertonic; shrivel; isotonic
7. 7.2 (% w/v)
8. Brownian motion
9. solute
10. $CCl_4$; $CH_3OH$; non-polar; non-polar
11. solvent (molecules); dilute; (more) concentrated
12. nephrons

Hofstadter's Law:
It always takes longer than you expect, even
when you take into account Hofstadter's Law.
- Douglas Hofstadter

# 7

# Reaction Rates and Equilibrium

## CHAPTER OBJECTIVES

After you have studied the chapter and worked the assigned exercises in the text and the study guide, you should be able to do the following.

1. Given the concentration of the reactants and/or products at various times during the reaction, calculate the rate of reaction.
2. Distinguish between molecular collisions and effective molecular collisions and how these relate to the rate of a reaction.
3. Clarify how effective collisions, temperature and activation energy relate to the rate of reaction.
4. Draw and label completely the energy diagrams for (i) an endothermic reaction and (ii) an exothermic reaction.
5. Describe how a catalyst influences a reaction and its affect on the energy diagram.
6. Define the composition of the activated complex and its position in the energy diagram.
7. Discuss how each of the following factors affect the rate of reaction: (1) nature of reactants, (2) concentration of reactants and products, (3) temperature and (4) the presence of a catalyst.
8. Describe the concept of dynamic equilibrium as it relates to a reversible reaction.
9. Given the equilibrium expression, write out the chemical reaction.
10. Given the chemical reaction, write out the equilibrium expression.
11. Given the concentration of the reactants and products at equilibrium, calculate the equilibrium constant, K.

12. Use LeChatelier's Principle to explain how the following factors influence the equilibrium: (1) addition of a reaction component; (2) removal of a reaction component; (3) addition of a catalyst.

13. Utilizing LeChatelier's Principle and considering both an exothermic and endothermic reaction, explain how a change in temperature (increase or decrease) will alter the equilibrium constant.

14. Define the important terms and comparisons in this chapter and give specific examples where appropriate.

## IMPORTANT TERMS AND COMPARISONS

Chemical Kinetics
Reaction Rate
Effective and Ineffective Collision
Activation Energy
Transition State and Activated Complex
Energy Diagram
Endothermic and Exothermic Reaction
Catalyst and Enzyme
Reversible and Irreversible Reaction
Dynamic Equilibrium
Haber Process
Fixation of Nitrogen
Equilibrium Constant and Equilibrium Expression

## SELF-TEST QUESTIONS

MULTIPLE CHOICE. In the following exercises, select the correct answer from the choices listed. In some cases, more than one choice will be correct.

1. Consider the reaction, $A + B \rightarrow C$. It was found that at 18 minutes after the start of the reaction, the concentration of C was 0.04 M. The rate of this reaction is:

   a. 450 M/min
   b. $4.2 \times 10^{-3}$ M/min
   c. 900 mole/L-min
   d. $2.1 \times 10^{-3}$ mole/L-min

2. Molecular collisions in the reaction, $A + B \rightarrow C$, are necessary for a reaction to take place because:

   a. Reactant molecules must directly interact with each other to produce product.
   b. The collisions of rapidly moving molecules provide energy.
   c. Both reasons above.

3. The activated complex in the reaction $A + B \rightarrow C + D$ is composed of:

   a. A and B
   b. C and D
   c. A
   d. A and D

4. Reaction $X + Y \rightarrow Z$ is completed in 2 hours at 20°C. The reaction is carried out at a different temperature at which the reaction is complete in 7.5 minutes. This new temperature is:

   a. 293 K
   b. 40°C
   c. 333 K
   d. 30°C

5. The equilibrium constant is equal to $5x10^{-4}$ for the reaction A + B + C = 2D. The equilibrium expression for the reaction is:

a. $K = \frac{(A)\ (B)\ (C)}{(D)^2}$

b. $K = \frac{(2D)}{(A)\ (B)\ (C)}$

c. $K = \frac{(D)^2}{(A)\ (B)\ (C)}$

d. $K = \frac{(2D)^2}{(A)\ (B)\ (C)}$

6. If the equilibrium concentrations for the reactants and products in the reaction in question 5 are [A] = [B] = [C] = $1x10^{-3}$ M, the concentration of D would be:

a. $5x10^{-10}$ M
b. $5x10^{-13}$ M
c. $7x10^{-7}$ M
d. $5x10^{-5}$ M

7. For the reaction $Cl_{2(g)} + H_2O_{(l)} = H^+ + Cl^- + HOCl$, the molarity of the chemical species at equilibrium are determined to be: $[Cl_{2(g)}] = 0.10$; $[H_2O] = 0.24$; $[H^+] = 1x10^{-3} = [Cl^-]$; $[HOCl] = 0.4$. The equilibrium constant for this reaction is:

a. $1.7x10^{-5}$
b. $1.7x10^{-2}$
c. $6x10^4$
d. $6x10^2$

8. Which of the following describe the Haber process?

a. $N_{2(g)} + 3H_{2(g)} = 2NH_{3(g)}$
b. Exothermic
c. Uses a catalyst
d. All of the above

9. If one monitors the rate of the formation of C in the reaction, A + B → C, one finds the rate is $5x10^{-2}$ M/min. After one hour of reaction, the concentration of C is:

a. $5x10^{-2}$ M
b. 3 M
c. $8x10^{-4}$ M
d. 30 M

COMPLETION. Write the word, phrase or number in the blank space which completes the statement.

1. Indicate whether the following reactions are endothermic or exothermic.

a. A + B + Energy → C ______________________

b. A + B → C + D + Energy ______________________

c. A + B → C
$\Delta E = -40$ kcal/mole ______________________

d. $H_{2(g)} + O_{2(g)} \rightarrow H_2O_{(e)}$ + Energy ______________________

2. Indicate how each of the following will affect the rate of reaction, A + B = C. The answer should be increase, decrease or no effect (on rate).

   a. Add a catalyst ______________________
   b. Increase concentration of A ______________________
   c. Decrease concentration of B ______________________
   d. Decrease the temperature ______________________

3. Consider the exothermic reaction, $A + B \rightarrow C + D$, -10 kcal, in which the activation energy for the reaction is 25 kcal. <u>Complete</u> the diagram below with labels and showing all items of importance.

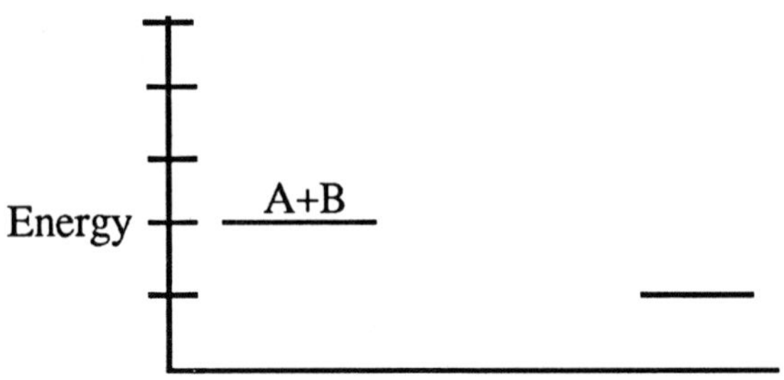


4. Completely label the following diagram by placing the appropriate label in each indicated block.

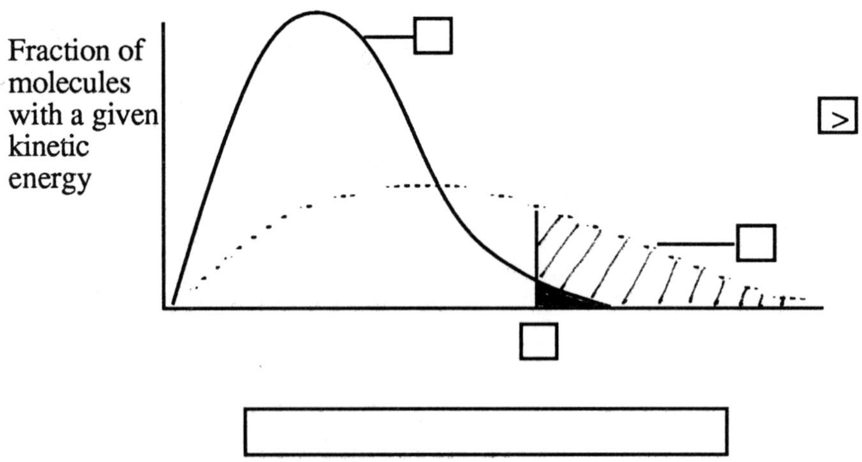


5. Molecules which act as catalysts in our bodies are called ________________.

6. Catalysts in a reaction lower the ________________ of the reaction by increasing the number of ________________ between reactant molecules without increasing the reaction temperature. However, the energy of the reactants and products <u>(do or do not)</u> ____________ change by the presence of a catalyst.

7. Complete the following table.

| <u>Equilibrium Expression</u> | <u>Reaction</u> |
|---|---|
| a. | $H_2 + O_2 = H_2O$ |

b. $\frac{(W)^2}{(A)(B)} = K$

c. $H_2O = H^+ + OH^-$

8. The concept of a dynamic equilibrium brings out the finding that although the _______________ of both reactant and product molecules (does or does not) change, both the _______________ and _______________ reactions are taking place.

9. Using LeChatelier's Principle, indicate how the following changes will affect the equilibrium in the exothermic reaction. Indicate whether the reaction goes to the right or to the left or has no change.

$$A + B = C$$

| Change | Effect |
|---|---|
| a. Increase [A] | |
| b. Decrease [C] | |
| c. Add a catalyst | |
| d. Increase temperature | |

## ANSWERS TO SELF-TEST QUESTIONS

### Multiple Choice

1. d
2. c
3. a and b; note that A and B are the only reactants and C and D are the only products. The atoms in A and B are the same ones in C and D, only rearranged to form the new molecules.
4. c
5. c
6. c
7. a
8. d
9. b

### Completion

1. a. endothermic
   b. exothermic
   c. exothermic
   d. exothermic

2. a. increase rate
   b. increase rate
   c. decrease rate
   d. decrease rate (number of effective collisions is reduced)

3.

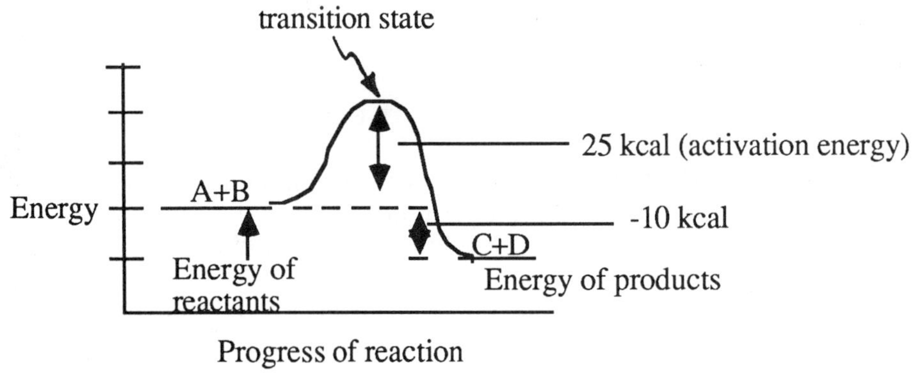


4.

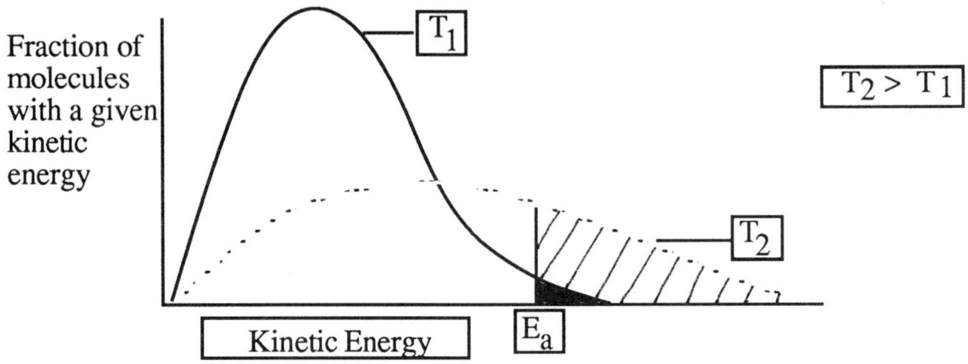


5. enzymes

6. activation energy, effective collisions, do not

7. a. $K = \frac{(H_2O)}{(H_2)(O_2)}$

   b. A + B = 2 W

   c. $K = \frac{(H^+)\ (OH^-)}{(H_2O)}$

8. concentrations, does not, forward, reverse

9. a. reaction goes to right
   b. reaction goes to right
   c. no effect
   d. reaction goes to left. In an exothermic reaction, heat can be considered a product. Therefore, increasing the temperature will increase the heat (a product) and the reaction will go to the left to relieve the stress.

Concerning solutions to acid-base
equilibrium problems: The answers
to the riddle of life are not found
in the back of the book.
- Charlie Brown

# 8

# Acids and Bases

## CHAPTER OBJECTIVES

As will become evident in this chapter and those focused on biochemistry (Chapters 16-26), the body contains aqueous systems vital to life as we know it. These systems include intracellular fluids, blood, gastric juices, saliva and more. The importance of understanding the **CHARACTERISTICS AND REACTIONS** of **ACIDS, BASES** and **SALTS** in aqueous solutions cannot therefore be overemphasized. This chapter dwells on these fundamentals.

After you have studied the chapter and worked the assigned exercises in the text and the study guide, you should be able to do the following.

1. Characterize Arrhenius acids and bases and define the similarities and differences with Bronsted-Lowry acids and bases. Give specific examples of each.

2. Explain the difference between strong and weak acids (bases) and give four examples of strong acids (bases).

3. Write the reactions involving Bronsted-Lowry acids and bases and point out the conjugate acid-base pairs.

4. Define Ka and pKa values and explain how they are related to acid strength.

5. Write ionic equations for the neutralization reaction of acids and bases; clearly define the salt which forms in each case.

6. For an aqueous solution, define the neutral point and the acidic and basic domains in terms of pH, pOH, $[H^+]$ and $[OH^-]$. At any pH, show the relationship between pH and pOH and also $[H^+]$ and $[OH^-]$.

7. Given the $[H^+]$ or $[OH^-]$ of a solution, calculate the pH. Given the pH or pOH of a solution, calculate the $[H^+]$.

8. Name six body fluids or common materials which are (1) acidic or which are (ii) basic. In each case, indicate its approximate pH value.

9. Explain the importance of acid-base indicators in qualitatively testing the acidic or basic character of a solution.

10. Make a list of salts and characterize them as either neutral, acidic or basic salts.

11. Write equations for the hydrolysis reactions associated with acidic and basic salts. Indicate whether the resulting solution is acidic or basic.

12. Define buffer solution and indicate the components and their relative concentrations necessary to prepare buffers with good buffering capacity.

13. Show how the components of carbonate and phosphate buffers react on addition of strong acid or base.

14. Explain in some detail the way in which you would determine the concentration of a basic solution by titration with a strong acid of known concentration.

15. Name a few acids and bases in which the equivalent weight is (i) the same as the formula weight, (ii) different than the formula weight. Explain the reason for this difference.

16. For the same acids and bases as in objective 15, indicate in which cases the normality of the solution is equal to or greater than the molarity.

17. Using the formula, $V_{acid} \times N_{acid} = V_{base} \times N_{base}$, and the data from a titration, calculate the normality of an acidic or basic solution of unknown concentration.

18. Define the important terms and comparisons in this chapter and give specific examples where appropriate.

## IMPORTANT TERMS AND COMPARISONS

Arrhenius Acid and Base
Hydronium Ion and Hydroxide Ion
Strong and Weak Acid
Glacial Acetic Acid
Bronsted-Lowry Acid and Base
Conjugate Acid-Base Pair
Mono-, Di- and Triprotic Acid
Acidity Constant, Ka
pKa
Ionic Equation
Antacid
Metal Oxide and Hydroxide
Baking Soda
Amine
Ion Product of Water, Kw
Acidic and Basic Solution
pH Indicator
pH Paper
pH Meter
Hydrolysis
Basic, Acidic or Neutral Salts
Buffering Action
Acidosis and Alkalosis
Hyper- and Hypoventilation
Buret
Pipet
Molecular Weight and Equivalent Weight
Molarity and Normality

## SELF-TEST QUESTIONS

MULTIPLE CHOICE. In the following exercises, select the correct answer from the choices listed. In some cases, more than one choice will be correct.

1. The pH of a $1 \times 10^{-4}$ M NaOH solution is

a. 4
b. 1
c. 10
d. 6

2. Which of the following is (are) strong acid(s)?

a. perchloric acid
b. boric acid
c. hydroiodic acid
d. acetic acid

3. Which of the following is both a strong acid and an oxidizing agent?

a. $NH_3$
b. $H_2SO_4$
c. $HNO_3$
d. $H_2O$

4. Indicate which of the following characteristics are associated with a strong acid solution.

a. concentrated
b. low pH value
c. dissociates close to 100%
d. reacts with bases

5. The species in a phosphate buffer at pH 6.5 which reacts with added hydroxide ion is:

a. $H_2PO_4^-$
b. $H_3PO_4$
c. $HPO_4^{-2}$
d. $PO_4^{-3}$

6. The formula weight of a triprotic acid is 144. The equivalent weight is:

a. 432
b. 72
c. 48
d. 288

7. The Ka value for lactic acid is $1.4 \times 10^{-4}$. The pKa value is:

a. 4.1
b. 3.8
c. 4.6
d. 3.1

8. A solution with pH = 7.7 has a $[H^+]$ of:

a. $2 \times 10^{-7}$ M
b. $7 \times 10^{-7}$ M
c. $2 \times 10^{-8}$ M
d. $7 \times 10^{-8}$ M

9. The pKa values for four acids are 2.6, 3.6, 4.9 and 5.2. The weakest acid is the one with pKa value of:

a. 2.6
b. 3.6
c. 4.9
d. 5.2

10. An aqueous solution made up of 0.1 M sodium acetate ($NaCH_3CO_2$) will be:

a. acidic
b. basic
c. neutral
d. cannot determine

11. The pH of a solution of pII 8.0 does not change upon addition of strong acid. The original solution:

a. contains a strong acid
b. contains an indicator
c. is a buffer solution
d. tastes sour

12. The normality of 100 mL of 0.4 M $H_2SO_4$ solution is:

a. 0.8
b. 0.4
c. 0.2
d. 0.04

13. The normality of a solution made up of $2 \times 10^{-4}$ moles of $Mg(OH)_2$ in 400 mL of water is:

a. $1 \times 10^{-6}$ N
b. $5 \times 10^{-5}$ N
c. $1 \times 10^{-3}$ N
d. $5 \times 10^{-6}$ N

14. Considering the acids below, together with the corresponding conjugate bases, which buffer system would be used to produce a buffer solution at pH 9.0?

a. acetic acid, $K_a = 1.8 \times 10^{-5}$
b. boric acid, $K_a = 7.3 \times 10^{-10}$
c. formic acid, $K_a = 1.8 \times 10^{-4}$
d. carbonic acid, $K_a = 4.3 \times 10^{-7}$

15. The Henderson-Hasselbach equation is used to calculate which of the following:
a. whether an acid is a strong or weak acid
b. whether a reaction between an acid and base will go to completion
c. the extent of acidosis or alkalosis in the blood
d. the ratio of the concentrations of the acid and its conjugate base to be used in preparing a buffer solution at the desired pH.

COMPLETION. Write the word, phrase or number in the blank space in answering the question.

1. Complete and balance the following reactions.

a. $HCl + NaOH \rightarrow$ ________ + ________ + ________

b. $Na_2O + HNO_3 \rightarrow$ ________ + ________ + ________

c. $H_2SO_4$ + ________ $\rightarrow K_2SO_4 + 2H_2O$

d. $NH_3 + HCl \rightarrow Cl^- +$ ________

e. $Na_2CO_3 + HI \rightarrow$ ________ + ________ + ________
final products

2. A 100 mL sample of unknown acid is titrated with 0.42 N NaOH. The end-point of the titration requires 29 mL of base. The normality of the acid is ________________.

3. If the acid in question 2 were $H_2SO_4$, the molarity of the acid would be ________________.

4. Indicate (yes or no) whether the following bases are Arrhenius and/or Bronsted-Lowry bases.

| | | Arrhenius | Bronsted-Lowry |
|---|---|---|---|
| a. | $NH_3$ | | |
| b. | NaOH | | |
| c. | $Al(OH)_3$ | | |

5. The term hydronium ion is used interchangeably with __________ , ____________ and ______________.

6. The ____________ of an acid in an intrinsic property of an acid, while the concentration is a variable and depends on the preparation of the solution.

7. Indicate the conjugate acid or base to the species below.

   a. $H_3O^+$ ______________________ (conjugate base)

   b. $H_2O$ ______________________ (conjugate base)

   c. $HSO_4^-$ ______________________ (conjugate acid)

   d. $Cl^-$ ______________________ (conjugate acid)

8. Indicate whether the acids shown below are mono-, di- or triprotic acids.

   a. $H_3PO_4$ ______________________

   b. $H_3CCOOH$ ______________________

   c. $H_2S$ ______________________

   d. $H_2CO_3$ ______________________

   e. $H_2PO_4^-$ ______________________

9. Complete the following table

| | $[H^+]$ | [OH] | pH |
|---|---|---|---|
| a. | $1 \times 10^{-2}$ M | ____________________ | _____ |
| b. | ________ | $2.5 \times 10^{-3}$ M | _____ |
| c. | ________ | ____________________ | 6.4 |

10. Indicate whether the dissolution of the following salts will produce a neutral, acidic or basic solution.

    a. KCl ______________________

b. $NH_4I$ ______________________

c. Na $(HCO_2)$ (formic acid is a weak acid) ______________________

d. $NaClO_4$ ______________________

11. If the pH of the blood goes below the normal pH, the condition is called ______________. This may be caused by ____________________________, a difficulty in breathing.

## ANSWERS TO SELF-TEST QUESTIONS

Multiple Choice

1. c
2. a, c
3. b, c
4. c, d
5. a
6. c
7. b
8. c
9. d
10. b
11. c
12. a (note that the volume has no influence on normality)
13. c
14. b
15. d

Completion

1. a. $H_2O$, $Na^+$, $Cl^-$
   b. $H_2O$, 2 $Na^+$, 2 $NO_3^-$
   c. 2 KOH
   d. $NH_4^+$
   e. $H_2O$, $CO_2\uparrow$, 2 NaI

2. 1.4 N

3. 0.7 M

4. a. no, yes
   b. yes, yes
   c. yes, yes

5. proton, $H^+$, $H_3O^+$

6. strength

7. a. $H_2O$
   b. $OH^-$
   c. $H_2SO_4$
   d. HCl

8. a. triprotic
   b. monoprotic
   c. diprotic
   d. diprotic
   e. diprotic

9. a. $1 \times 10^{-12}$ M, 2

   b. $4.0 \times 10^{-12}$, 11.4

c. $4 \times 10^{-7}$, $2.5 \times 10^{-8}$

10. a. neutral; salt of a strong acid (HCl) and a strong base (KOH)
    b. acidic, salt of a strong acid (HI) and a weak base ($NH_4OH$)
    c. basic; salt of a weak acid ($HCO_2H$) and a strong base (NaOH)
    d. neutral; salt of a strong acid ($HClO_4$) and a strong base (NaOH)

11. acidosis, hypoventilation

Every man, woman and child inhabiting this planet
became an unwitting guinea pig in a vast
biochemical experiment on July 16, 1945, when
the first atomic bomb was successfully detonated.
- Ernest Borek
The Atoms Within Us

# Nuclear Chemistry

## CHAPTER OBJECTIVES

After you have studied the chapter and worked the assigned exercises in the text and the study guide, you should be able to do the following.

1. Characterize the three major types of radioactivity in terms of symbol, charge, mass and penetrating capability.
2. Associate the names of famous scientists with their accomplishment in nuclear chemistry.
3. Given the frequency ($\upsilon$) of electromagnetic radiation and the equation ($\lambda$) ($\upsilon$) = c, calculate the wavelength ($\lambda$). Given the wavelength, calculate the frequency.
4. Order the different forms of radiation in the electromagnetic spectrum in terms of increasing energy and frequency and decreasing wavelength.
5. Explain the difference between a chemical and a nuclear reaction.
6. Write balanced nuclear equations for a reaction involving an (i) alpha particle, (ii) beta particle and (iii) gamma emission. Note whether the product nuclei has a mass greater than, equal to or less than the original radioactive nuclide.
7. Given an equation which includes all but one particle in a nuclear reaction, balance the equation and identify the missing particle.
8. Explain the term "half-life" and relate this to the decay curve for a radioactive nuclide.
9. Explain how a Geiger-Muller counter functions in monitoring the intensity of beta particle emission.

10. Given the total amount of radioactivity (mCi) in a given volume of sample, calculate the amount of radioactivity in small aliquots of the sample.

11. Define the three units, roentgens, rems and rads, used to describe the effects of radiation and indicate how they relate to each other.

12. Describe the role of cobalt-60, iodine-131, and technetium-99m in medical diagnosis and/or treatment.

13. Explain the importance of fusion reactions to life on earth.

14. Point out the most obvious differences between nuclear fusion and fission reactions. Outline how the fission reactions can be used to produce controlled nuclear energy on the one hand and the atomic bomb on the other hand.

15. Define the important terms and comparisons in this chapter and give specific examples where appropriate.

## IMPORTANT TERMS AND COMPARISONS

Alpha ($\alpha$) Particle, Beta ($\beta$) Particle, and Gamma ($\gamma$) Ray
Wavelength ($\lambda$) and Frequency ($\upsilon$)
$(\lambda)(\upsilon) = c$
Photon and electron
Electromagnetic Spectrum
Subatomic Particle
Nuclide
Mass Number and Atomic Number
Hydrogen, Deuterium and Tritium
Isotope
Alpha, Beta or Gamma Emitter
Natural and Artificial Transmutation
Half-Life ($T_{1/2}$)
Radioactive Decay Curve
"Phosphor"
Geiger-Muller Counter
Spectrum of Radiation
Roentgen, Rad and Rem
Equivalent Dose
Exposure Badge
Mutation
Background Radiation
Hyperthyroidism
Tagging a Drug
Collimated Radiation Beam
Radiation Sickness
"Big Bang"
Fusion
$E = mc^2$
Isotope Generator
Nuclear Fission and Fusion
Nuclear Power Plant
Control Rod
Nuclear Waste
Breeder Reactor

## SELF-TEST QUESTIONS

MULTIPLE CHOICE. In the following exercises, select the correct answer from the choices listed. In some cases, more than one choice will be correct.

1. A 4 g sample of a radionuclide at room temperature has a half-life of three months. The amount of sample remaining after a year is:

   a. 0.25 g
   b. 0.40 g
   c. 2 g
   d. 4 g

2. If the radioactive decay of the sample in exercise 1 were carried out -100°C, the decay process would:

a. decrease
b. increase
c. remain unchanged
d. cannot determine

3. A nucleus which emits a beta particle will be transformed into a product nucleus of the:

a. same mass number
b. same atomic number
c. atomic number one unit greater
d. mass number one unit greater

4. The wavelength of blue light is 4000 A (1 A = $10^{-8}$ cm). Its frequency is:

a. $12x10^{-12}$ $s^{-1}$
b. $7.5x10^{14}$ $s^{-1}$
c. $7.5x10^{10}$ $s^{-1}$
d. $12x10^{10}$ $s^{-1}$

5. A scintillation counter uses a "phosphor" to:

a. determine $t_{1/2}$ values
b. produce transuranium elements
c. measure the number of mutations
d. measure the radioactivity in a sample

6. A millicurie is a unit of radioactivity equivalent to:

a. $3.7x10^{10}$ counts $s^{-1}$
b. $3.7x10^{10}$ counts $min^{-1}$
c. $3.7x10^{7}$ counts $s^{-1}$
d. $3.7x10^{4}$ counts $s^{-1}$

7. The radiation which causes the most damage within the cell is:

a. alpha particles
b. beta particles
c. gamma rays
d. X-rays

8. Which of the following radioactive nuclides decay the fastest?

a. $^{197}_{80}Hg$, $t_{1/2}$ = 65 hrs
b. $^{210}_{84}Po$, $t_{1/2}$ = 138 days
c. $^{238}_{92}U$, $t_{1/2}$ = $4x10^9$ yr.
d. $^{131}_{53}I$, $t_{1/2}$ = 8 days

9. The wavelength of which radiation is the greatest (longest)?

a. microwaves
b. X-rays
c. visible light
d. radiowaves

10. The number of neutrons in phosphorus-32, $^{32}_{15}P$ is:

a. 32
b. 15
c. 17
d. 5

11. Which of the following nuclear reactions does not involve natural transmutation?

a. $^{11}_{5}B \rightarrow {}^{11}_{5}B + \gamma$

b. $^{210}_{84}Po \rightarrow {}^{206}_{82}Pb + {}^{4}_{2}He$

c. $^{14}_{6}C \rightarrow {}^{14}_{7}N + {}^{0}_{-1}e$

COMPLETION. Write the word, phrase or number in the blank space in answering the question.

1. Consider the reactions in question 11 in the multiple choice section. Reaction a is an example of a(n) ____________ emitter, reaction b shows a(n) ____________ emitter, while reaction c exhibits a(n) ____________ emitter.

2. Complete the following nuclear reactions. Refer to a periodic table.

a. $^{28}_{13}Al \rightarrow \beta +$ ____________

b. ____________ $\rightarrow {}^{1}_{1}H + {}^{0}_{-1}e$

c. $^{254}_{102}No \rightarrow {}^{4}_{2}He +$ ____________

d. $^{87}_{36}Kr \rightarrow {}^{87}_{37}Rb +$ ____________

3. The order of increasing penetrating power for gamma rays, alpha and beta particles is

____________ < ____________ < ____________

4. Match up the names of the famous scientist in the left-hand column with accomplishments or subjects most associated with the individual.

| | | | |
|---|---|---|---|
| a. | H. Becquerel | (i) | Discovered X-ray |
| b. | E. Rutherford | (ii) | $E = mc^2$ |
| c. | G. Seaborg | (iii) | CAT scan |
| d. | A. Einstein | (iv) | Prepared transuranium elements |
| e. | W. Roentgen | (v) | "Coined" the names alpha, beta and gamma rays |
| f. | A. Cormack and G. Hounsfield | (vi) | $3.7 \times 10^{10}$ counts $s^{-1}$ |
| g. | M. Curie | (vii) | Observed that certain rocks gave off mysterious radiation |

5. Radiation that is derived from both cosmic rays and rocks in our environment contribute to the so-called ____________ radiation.

6. Another name commonly used for an alpha particle is the ____________ of a ____________ atom.

7. A beta particle has mass number of _______________ and an atomic number of _______________.

8. Cosmic rays are streams of _______________, while gamma rays consist of _______________, _______________.

9. Isotopes have the same number of _______________ but different number of _______________.

10. A sample of a radioactive nuclide with a half life of 10 years is left in a cave for 35 years. After this time, only 0.5 grams remains. The amount of this radioactive material originally left in the cave was _______________.

11. A sample of radioactive isotope has an activity of 4.0 mCi. This sample emits _______________ counts per second.

12. Diagnostic X-rays are absorbed by the bones and not the _______________ because the bones contain elements of _______________ atomic number which absorb X-rays strongly.

13. Consider the following two reactions:

    a. $^{252}_{98}Cf + ^{10}_{5}B \rightarrow ^{257}_{103}Lr + 5\ ^{1}_{0}n$

    b. $^{235}_{92}U + ^{1}_{0}n \rightarrow ^{90}_{37}Rb + ^{144}_{55}Cs + 2\ ^{1}_{0}n$

    Reaction (a) is an example of a _______________ reaction, while reaction (b) is a _______________ reaction.

14. An important characteristic of a breeder reaction is that it not only provides _______________, but also makes additional _______________.

## ANSWERS TO SELF-TEST QUESTIONS

### Multiple Choice

1. a
2. c
3. a, c
4. b
5. d
6. c
7. a
8. a
9. d
10. c
11. a

### Completion

1. a. gamma
   b. alpha
   c. beta

2. a. $^{28}_{14}Si$

b. ${}^{1}_{0}n$

c. ${}^{250}_{100}Fm$

d. ${}^{0}_{-1}e$

3. alpha < beta < gamma

4. a. (vii)
   b. (v)
   c. (iv)
   d. (ii)
   e. (i)
   f. (iii)
   g. (vi)

5. background

6. nucleus, helium

7. zero, minus one

8. nuclei (mainly protons), high energy radiation

9. protons, neutrons

10. ca. 6 g

11. $1.5 \times 10^8$

12. soft tissue, higher

13. fusion, fission

14. energy, fuel

Gasoline - 29¢/gal
- Sign at gas station in
Forty Fort, PA; circa 1965

# 10

# Organic Chemistry. Alkanes

## CHAPTER OBJECTIVES

After you have studied the chapter and worked the assigned exercises in the text and the study guide, you should be able to do the following.

1. Distinguish between the areas of organic and inorganic chemistry and characterize the properties of compounds which are representative of each.
2. Point out at least one feature which is common to organic and biochemistry and then note aspects in which they differ.
3. Characterize the different families of hydrocarbons and the types of (carbon-carbon) bonds that are found in each family.
4. Using the six IUPAC rules of nomenclature for saturated hydrocarbons, write the names for alkanes and alkyl halides.
5. Write out the structural formulas of molecules for which the IUPAC names are given.
6. Distinguish between the different types of functional isomers and stereoisomers.
7. Characterize the most important chemical and physical properties of alkanes.
8. Outline the most important characteristics of the $sp^3$ hybrid orbital.
9. Name the most important functional groups outlined in Table 10.7 and draw a molecule with each in it.
10. Define the important terms and comparisons in this chapter and give specific examples where appropriate.

## IMPORTANT TERMS AND COMPARISONS

Organic, Inorganic and Biochemistry
Biosynthesis
Molecular and Structural Formula
Isomeric Forms
Functional, Geometric and Sterioisomer
Cis- and Trans-Configuration
Alkane, Alkene and Alkyne
Primary, Secondary, Tertiary and Quaternary Carbon
Main Chain and Side Chain
Systematic and Common Names of Compounds
Di-, Tri-, Tetra-, Penta-, and Hexa-
Chair and Boat Shape
Equitorial and Axial Position
Combustion
Straight Chain, Branched Chain
Hybrid Orbital and $sp^3$ Orbital

## SELF-TEST QUESTIONS

MULTIPLE CHOICE. In the following exercises, select the correct answer from the choices listed. In some cases, more than one choice will be correct.

1. Which of the following properties characterize organic compounds?

   a. Soluble in water
   b. Burn in air
   c. Bonding is primarily covalent
   d. Most have high-melting points
   e. Contains carbon and other elements

2. The molecular formula, $C_3H_8O$, could represent the following class(es) of organic compound(s).

   a. alkane
   b. ether
   c. alkyl chloride
   d. amine
   e. alcohol
   f. mercaptan

3. The family of hydrocarbons with a C=C (double) bond is called:

   a. alkene
   b. alkyne
   c. ether
   d. amine

4. Which hydrocarbon has one $-CH_2$ unit more than in $CH_3CH_2CH_2CH_2CH_2CH_2CH_3$?

   a. butane
   b. octane
   c. propylene
   d. heptane

5. Which of the following structural formulas represent the same compound?

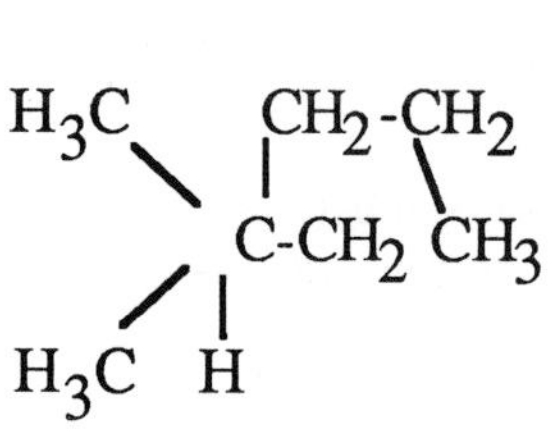

(i)

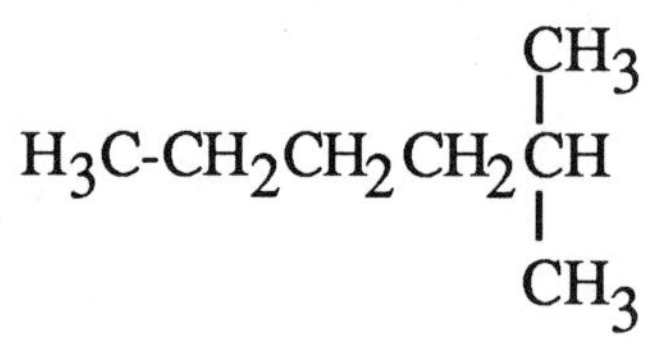

(ii)

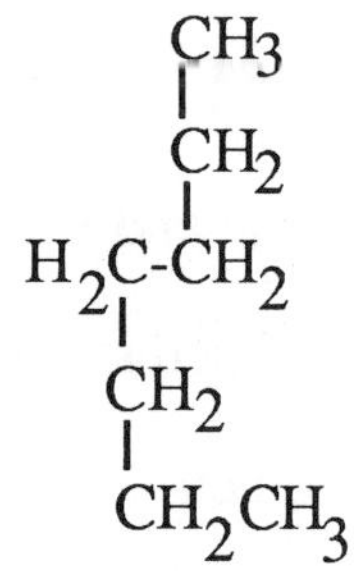

(iii)

a. all are different
b. (i) and (ii)
c. (i) and (iii)
d. all structural formulas

6. The IUPAC name for the compound below is:

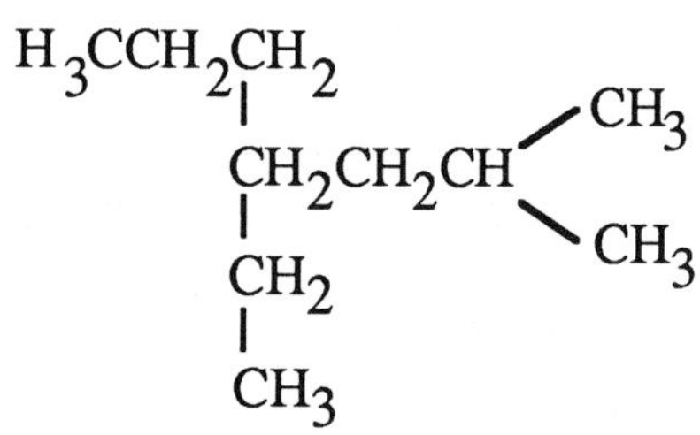

a. 2-methyl-4-propylhexane
b. 4-ethyl-7-methylheptane
c. 4- ethyl-2-methylheptane
d. n-decane

7. The structure for 5-bromo-2-chloroheptane is:

a. 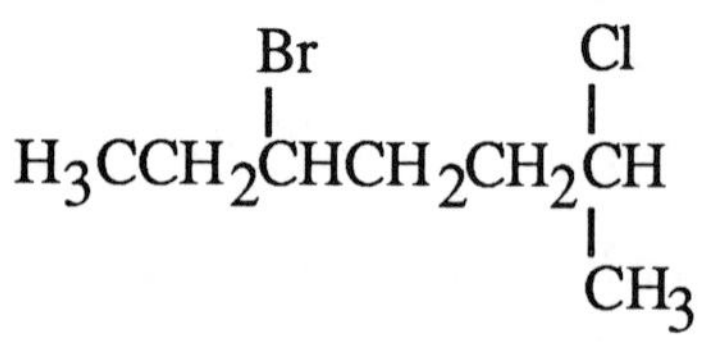

c. 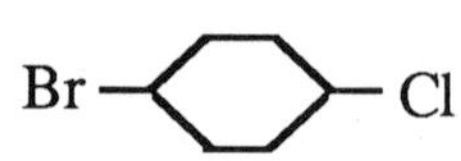

b. 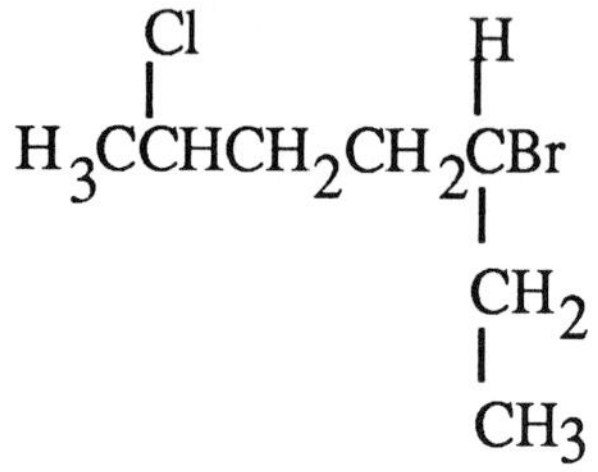

d. 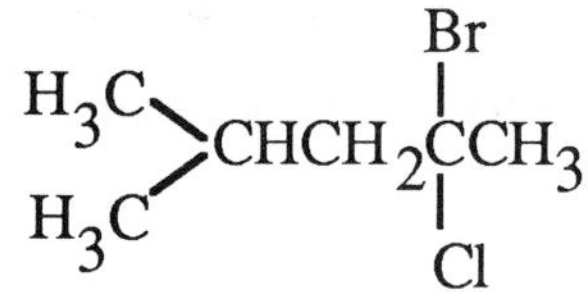

8. The combustion of pentane produces:

a. an alcohol
b. carbon dioxide and water
c. pentene
d. ethane

9. A tertiary carbon is bonded directly to:

a. 2 hydrogens
b. 3 carbons
c. 2 carbons
d. 4 carbons

10. The IUPAC name for 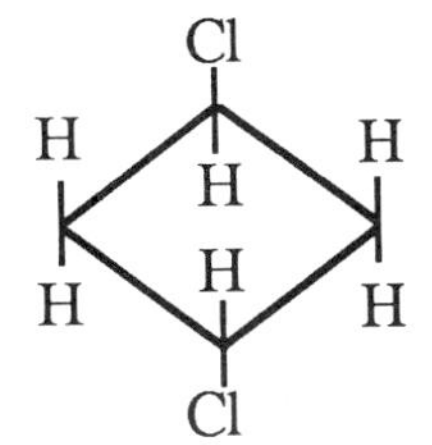 is:

a. trans-1,3-dichlorocyclobutane
b. cis-1,2-dichlorocyclobutane
c. cis-1,3-dichlorocyclobutane
d. cyclobutane

11. Which hydrocarbon molecule with eight carbons would produce the lowest octane number (i.e., the most knocking) when used as a fuel in an automobile?

a. 2-methyl-3,3-dimethylpentane
b. n-octane
c. 2-isopropylpentane
d. 2-methylheptane

12. The bonding in methane involves $sp^3$ hybrid orbitals. Which of the following is true about the orbital and the bonding?

a. each hybrid orbital has a 50% contribution from the s and the p orbital
b. each hybrid orbital is a mixture of one part s orbital and three parts p orbital
c. the angle between the hybrid orbital is 109.5 degrees.
d. there are four $sp^3$ hybrid orbitals formed

COMPLETION. Write the word, phrase or number in the blank space or draw the appropriate structure in answering the question.

1. Examine the following structural formulas and supply the information required by writing the identifying number(s) in the blank.

(i) $CH_3CH_2CH_2CH_2CH_3$

(ii) $H_3COCH_2CH_3$

(iii) 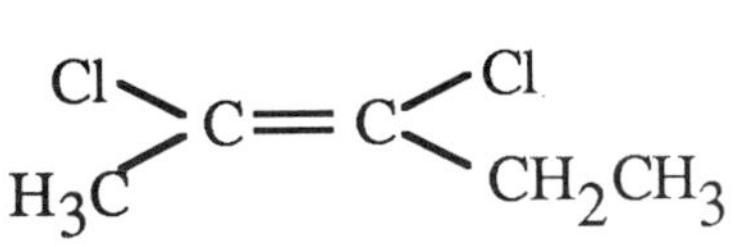

(iv) 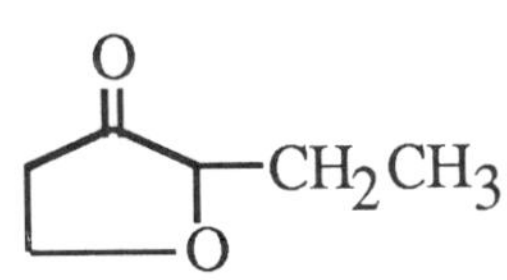

(v) 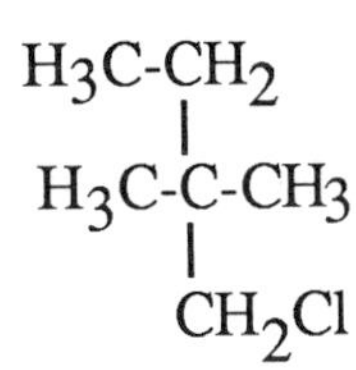

a. Hydrocarbon molecule ______________________
b. Contains ether functional group ______________________
c. Halogenated hydrocarbon ______________________
d. Unsaturated (C=C) bond ______________________
e. Straight chain molecule ______________________
f. Contains more than one functional group ______________________
g. Acyclic molecule ______________________

2. The outstanding characteristic of alkanes is their ______________________.

3. Isomers are defined as molecules which have the same ______________, but different ______________.

4. Ethanol which contains a small amount of ______________ is labeled ______________.

5. A simple ether molecule contains ______________ oxygen atom(s), while a carboxylic acid functional group contains ______________ oxygen atom(s).

6. The boiling point of a series of alkanes increases as the molecular weight of the alkane ______________.

7. Cyclohexane in the chair form has __________ hydrogens "pointing" up or down called __________ hydrogens, in addition to the ______________ hydrogens "pointing" approximately in the "plane" of the ring called __________ hydrogens.

8. Cis- and trans-isomers are examples of ______________ or __________ isomers.

9. Write the IUPAC name for the following molecules:

a. 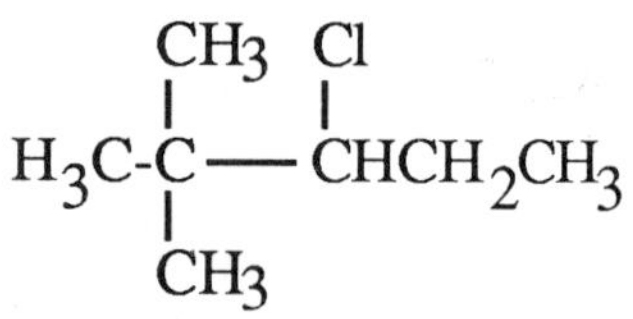 ______________________

b.  ______________________

c. 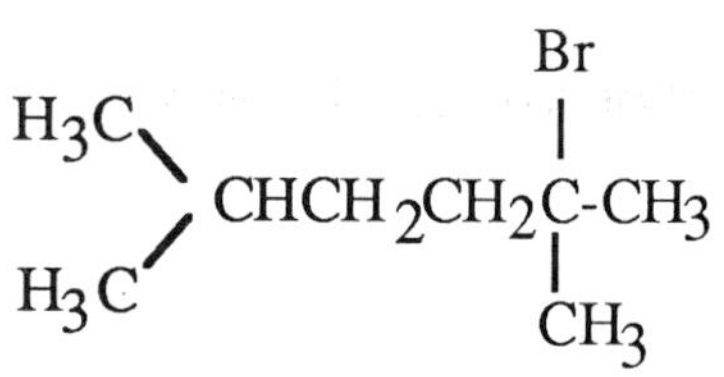 ______________________

10. Draw the structures for the following compounds, showing all the atoms.

a. 3-chloro-2-methylpentane ______________________

b. 1-iodo-3,3-dimethylcyclobutane ______________________

c. 4-isopropylheptane ______________________________

11. A saturated hydrocarbon can exist in a number of __________ because there is __________ rotation about a (C-C) single bond.

12. Give the family name for each functional group listed below.

a. R-C(=O)-H ______________________________

b. R-O-R' ______________________________

c. $R_2NH$ ______________________________

d. 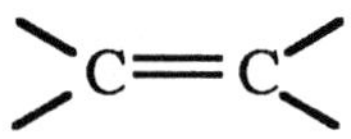 ______________________________

e. R-SH ______________________________

13. The structure of the cyclohexane ring is not _______________. The molecule exists in two conformations called the _______________ and the _______________ forms. In both of these forms, hydrogens pointing up-and-down are called _______________ hydrogens, while the others are referred to as _______________ hydrogens.

## ANSWERS TO SELF-TEST QUESTIONS

Multiple Choice

1. b, c, e
2. b, e
3. a
4. b
5. b
6. c
7. a and b are the same molecule
8. b
9. b
10. a
11. b
12. b, c, d

Completion

1. a. i only; all other molecules include C and H, in addition to other elements.
   b. ii, iv
   c. iii, v
   d. iii
   e. i, ii
   f. iv (and iii); III also, if alkyl halide is considered as a functional group
   g. i, ii, iii, v

2. lack of reactivity
3. molecular formula, structural formula
4. methanol, denatured alcohol

5. one, two
6. increases
7. 6, axial, 6, equatorial
8. stereoisomers, geometrical
9. a. 3-chloro-2,2-dimethylpentane
   b. 2-iodo-1-methylcyclopentane
   c. 2-bromo-2,5-dimethylhexane

10.

a. 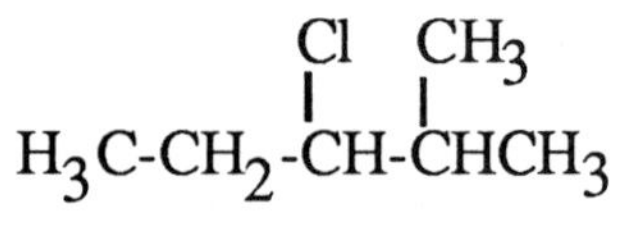

b. 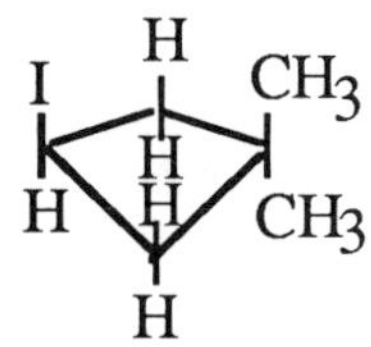

c. 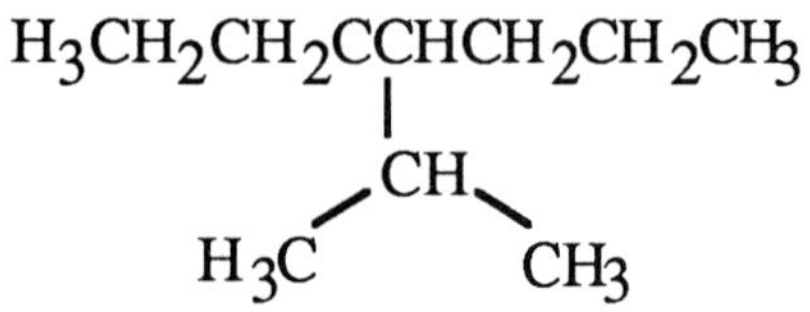

11. conformations, free
12. a. aldehydes
    b. ether
    c. amine
    d. alkene
    e. mercaptan or thiol
13. planar, boat, chair, axial, equatorial

On Chemistry:
It's hard and its hard, ain't it
hard, Good Lord
- Woody Guthrie
"Hard, Ain't It Hard," 1952

# 11

# Alkenes, Alkynes and Aromatic Compounds

## CHAPTER OBJECTIVES

Chapter 11 begins to present some of the substance of organic chemistry. This includes not only the formal naming of, but also the drawing of structures for organic molecules containing a particular functional group. In addition, there will be an increasing emphasis on recognizing the physical properties, and especially the chemical reactions, that can take place as a direct consequence of the presence of the functional group (i.e., the "reactive center"). The strategies and learning patterns developed in studying the material in this chapter will be especially useful also in subsequent chapters.

After you have studied the chapter and worked the assigned exercises in the text and the study guide, you should be able to do the following specific objectives.

1. Recognize and distinguish the structural differences in alkenes, alkynes and aromatic molecules.

2. Given the structure formulas for unsaturated hydrocarbons, name the molecules using the IUPAC rules.

3. Given the IUPAC name for an unsaturated compound, draw the structural formula which shows all the atoms.

4. Characterize the addition reactions in terms of the proposed intermediate and the products which occur as $H_2$, HCl and $Cl_2$ react with an alkene or alkyne.

5. Show how the orbital hybridization of carbon in alkenes and alkynes determines the geometry about the carbon atom.

6. Use Markovnikov's rule, as it applies to the addition of an unsymmetrical small molecule (HX, X = Cl, Br, OH) to an alkene or alkyne, to predict the reaction products.

7. Illustrate how specific alkene monomers can undergo polymerization reactions.

8. Draw the structure of the aromatic molecule, benzene, in its resonance hybrid forms (Lewis structures) and in the "loop form".

9. Characterize the substitution reactions involving nitration, sulfonation and halogenation which occur on aromatic molecules. Note which reactions require a catalyst.

10. Define the important terms and comparisons and give specific examples where appropriate.

## IMPORTANT TERMS AND COMPARISONS

Unsaturated and Saturated Hydrocarbon
Alkane, Alkene and Alkyne
Monomer and Polymer
IUPAC
Aromatic and Aliphatic Compound
Benzene
[$sp^2$ + p] and [sp + 2p] Hybridization

Reaction Types
a. Addition
b. Catalytic Hydrogenation
c. Hydration
d. Aromatic Substitution
e. Nitration
f. Sulfonation
g. Halogenation

Before challenging your comprehension of this material by working out the SELF-TEST, a few words are in order to help sort out the major themes presented. As this and subsequent chapters are individually addressed, it will become clear that the general objectives and the nature of the material is much the same. This is especially true in Chapters 12, 13, 14 and 15. Each of these chapters introduces one or more families of molecules with a characteristic functional group. **IT IS THE FUNCTIONAL GROUP WHICH IS THE REACTIVE CENTER OF THE MOLECULE.** That is, this is where the reaction will take place with other molecules. The functional group will have the major influence in defining the chemical and physical properties of a compound. Compare the chapter titles with the list of functional groups in Table 10.7 (text). The focus of Chapter 11 is aimed at basically three major aspects of the three families of unsaturated molecules. These three themes are:

1. The **naming** of unsaturated molecules using the internationally accepted IUPAC rules. This nomenclature systematically defines the composition and structure of the molecule.

2. The **characterization of the physical properties** of each family of molecules.

3. **Identification of the reactions** which characterize each type of unsaturated molecule.

These same themes will reoccur in the other chapters, with simply the focus reset on a different functional group.

Possibly the most troublesome area for students is developing a strategy in identifying and characterizing the reactions associated with a family of molecules. A guide or "road map" for the process, which is both simple and useful, is the following.

1. Identify the functional group.
2. Tabulate the reactions associated with the functional group.
3. Characterize the reaction by its general name (addition reaction, oxidation, etc.) and then tabulate any additional specifics (catalyst or heat required, follows Markovnikov's rule, etc.).

It is very helpful if the student can write out a condensed summary of the reactions for each family of molecules. An example for alkenes follows.

**Condensed Reaction Summary for Alkenes**

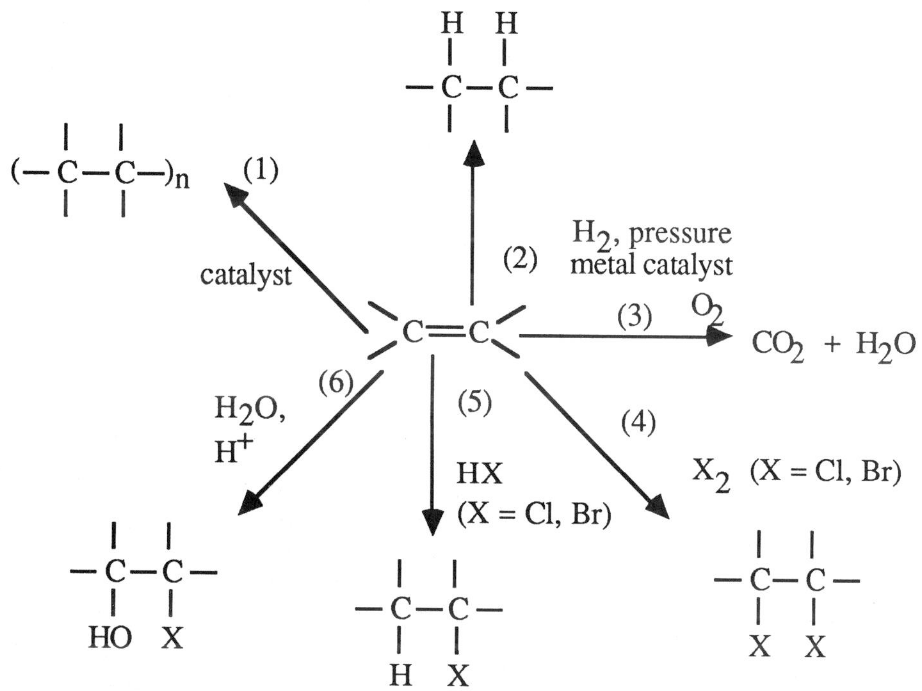


| **Reaction** | **Name** | **Comment** |
|---|---|---|
| (1) | Polymerization | Requires appropriate catalyst; addition reaction |
| (2) | Catalytic hydrogenation | Addition reaction; pressure and metal catalyst |
| (3) | Combustion or oxidation of a hydrocarbon | $CO_2$ and $H_2O$ are the common combustion products of all hydrocarbons |
| (4) | Bromination or chlorination | Addition reaction |
| (5) | Addition reaction of an HX gas | Both addition reactions (5) and (6) obey Markovnikov's rule |
| (6) | Hydration | |

## SELF-TEST QUESTIONS

MULTIPLE CHOICE. In the following exercises, select the correct answer from the choices listed. In some cases, two or more answers will be correct.

1. A terpene containing 20 carbons has the following characteristics:

   a. 4 isoprene units
   b. 4 double bonds
   c. the product of a plant
   d. contain alkyne units

2. The reaction of cyclopentene with $Br_2$ yields:

   a. 1,2-dibromopentene
   b. 1,2-dibromopentane
   c. 1,2-dibromocyclopentane
   d. 1,3-dibromocylcopentene

3. The carbocation intermediate formed in the reaction of HBr with 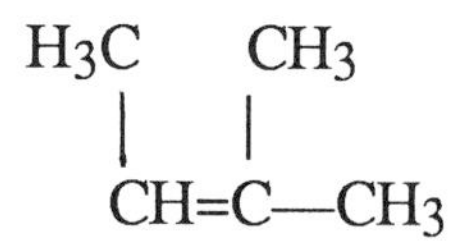 is:

   a. 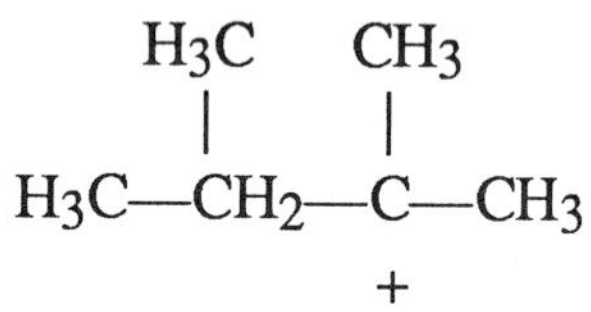

   c. 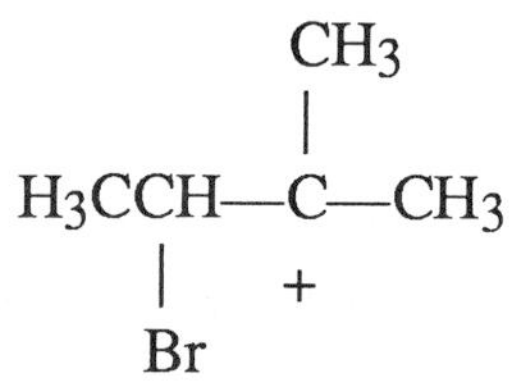

   b. 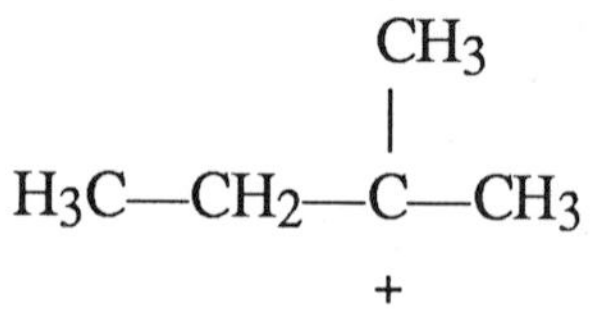

   d. 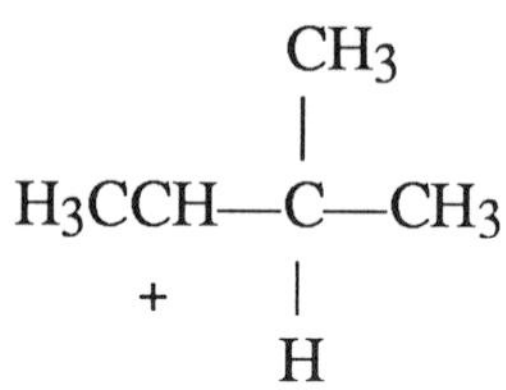

4. The principal product of the reaction of $Br_2$ with para-methyltoluene is:

   a. 2-bromo-1,4-dimethylbenzene
   b. 2,3-dibromo-1,4-dimethylbenzene
   c. no reaction
   d. 3-bromo-4-methyltoluene

5. Addition reactions do not occur with aniline because the ring:

a. is aromatic
b. is not unsaturated
c. already contains an $-NH_2$ group
d. is stabilized by resonance

6. The polymerization of vinyl chloride ($H_2C{=}CHCl$) produces:

a. unsaturated polyvinylchloride
b. saturated polyvinylchloride
c. a branched polymer
d. a linear polymer

7. An example of a secondary carbocation is:

a. $CH_3^+$

b. $H_3C{-}\overset{+}{C}H_2$

c. $(CH_3)_2C^+{-}CH_3$ (H3C—C+ with CH3 above and CH3 below)

d. $H_3C{-}\overset{+}{C}H{-}CH_3$

8. The structural formula for the cis-isomer of $ClHC{=}CCH_3Cl$ is:

a. 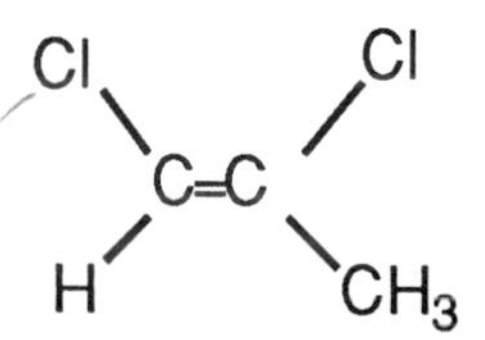

c. 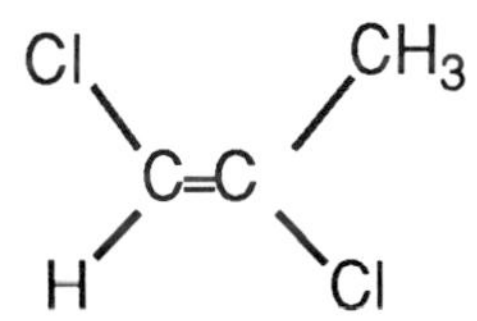

b. 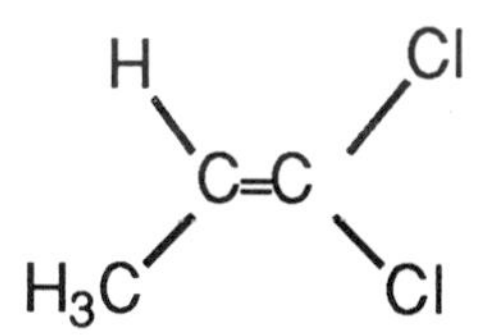

d. 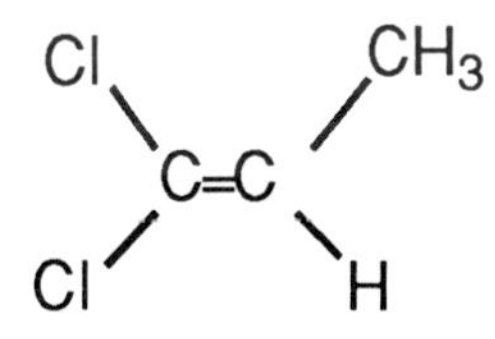

9. The reaction of HBr with which of the compounds shown below, will produce $H_3CCH_2CHBrCH_3$:

a. $H_2C{=}CHCH_2CH_3$
b. $H_3CCH{=}CHCH_3$
c. $H_3CCH_2CH_2CH{=}CH_2$
d. $H_3CC{\equiv}CCH_3$

10. The IUPAC name for the compound below is:

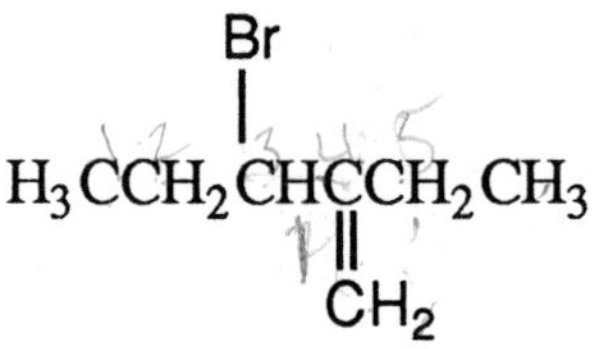

a. 3-bromo-2-ethyl,1-pentene
b. 3-bromo-2-ethyl,2-hexene
c. 3-bromo-4-ethyl,1-pentene
d. 3-bromo-4-ethyl,1-hexane

11. The name of the compound below is:

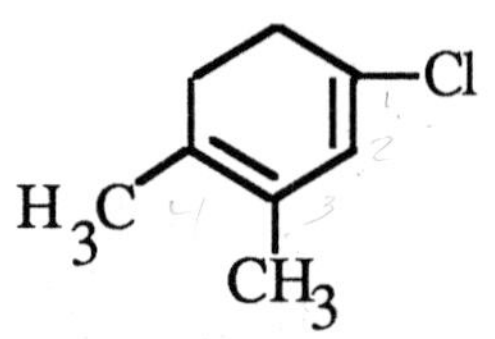

a. 1-chloro-3,4-dimethyl-2,3-cyclohexatriene
b. 1-chloro-3,4-dimethyl-1,3-cyclohexadiene
c. 3,4-dimethyl-1-chloro-1,3-cyclohexadiene
d. 1-chloroorthomethyl-1,4-cyclohexadiene

12. Which of the following compounds is not a heterocyclic compound?

a. 

c. 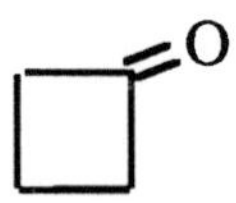

b. 

d. 

13. An $sp^2$ hybrid orbital in an organic compound has which of the following characteristics:

a. The $sp^2$ orbital is parallel to the p orbital
b. The $sp^2$ orbital is a sigma orbital
c. The $sp^2$ orbital has 25% s character
d. This type of orbital is always found in alkenes.

COMPLETION. Write the word, phrase or number in the blank or draw the appropriate structure in answering the question.

1. Give the IUPAC names for the following molecules:

a.  1,4 cyclohexdiene

b. 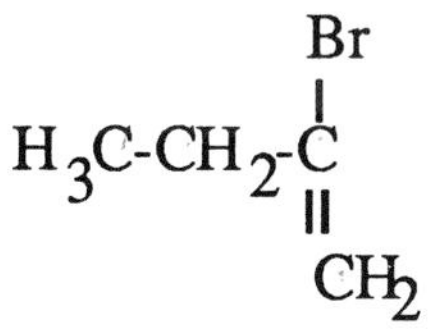 2 bromo 1 butene

c. 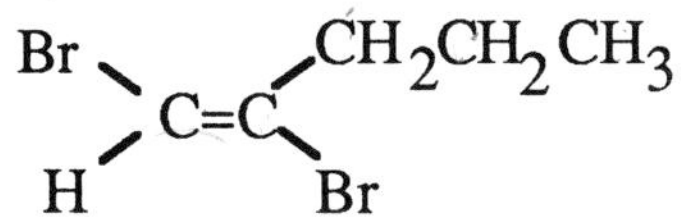 trans 1,2 bromo 1 pentene

d. 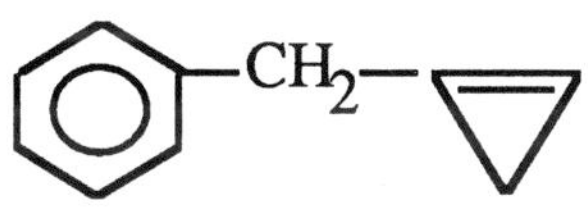 benzyl 1 cyclopropene

e. 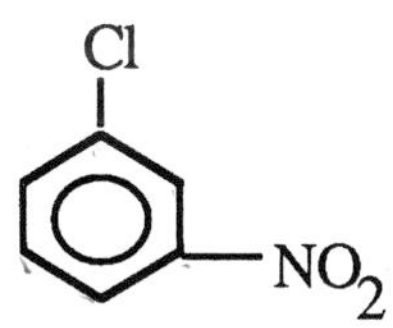 1,3 chloronitrate benzene

f. 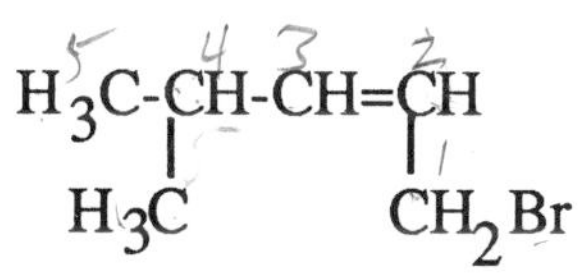 1 bromo 4 methyl 2 pentene

2. Sulfonation is defined as substitution or replacement of a H on a aromatic ring w/ the $-SO_3H$ group

3. Give the common names for the following molecules:

a. 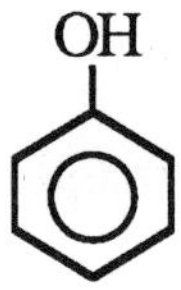 ________________________

b. $C_2H_2$ ________________________

c. $C_6H_6$ ________________________

d. 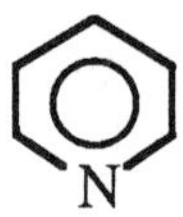 ________________________

e. 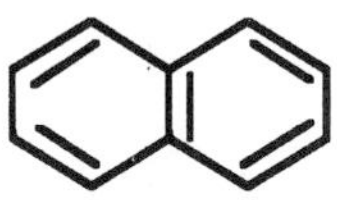 ________________________

f. $-[CH_2CH_2CH_2CH_2CH_2]-$ ________________________

4. Plastic combs are made of the polymeric material polymethylmethacrylate while indoor-outdoor carpeting is made of polypropylene. Draw the structural formulas for these products which are produced in the reactions below.

a. $H_2C{=}C(CH_3){-}C({=}O){-}OCH_3 + H_2C{=}C(CH_3){-}C({=}O){-}OCH_3 \xrightarrow{\text{catalyst}}$ ________________________

b. $H_2C{=}CHCH_3 + H_2C{=}CHCH_3 \xrightarrow{\text{catalyst}}$ ________________________

c. This polymerization reaction is also classified as an ________________________ reaction.

d. A copolymer is made up of ________________________.

5. The boiling point for 1-hexene is ______________ than that of 1-octene.

6. The carbons in acetylene contain ________ hybrid orbitals which have _______% s character. This molecule is __________ in structure and contains ____ pi orbitals. After hydrogenation of acetylene to produce ethylene, the carbons must rehybridize to form ________ hybid orbitals. This orbital contain s ________% s character and __________% p character.

7. Fill in the structural formula of the product in the following reactions and classify each equation according to the type of reaction occurring.

| | Product | Reaction Type |
|---|---|---|

a. $H_3CCHCH_2 + O_2 \longrightarrow$

b. 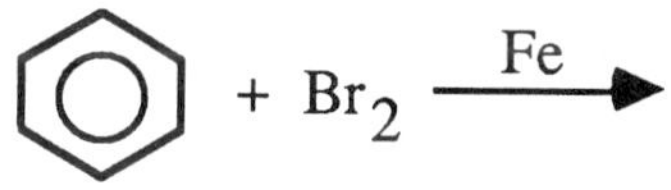


c. 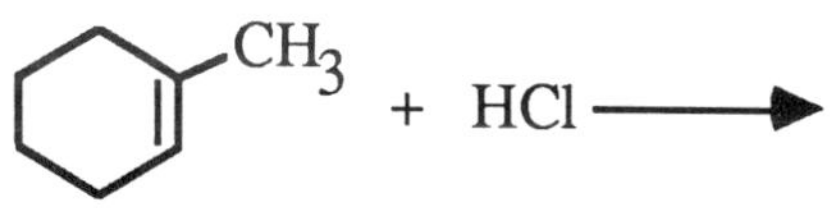


d. 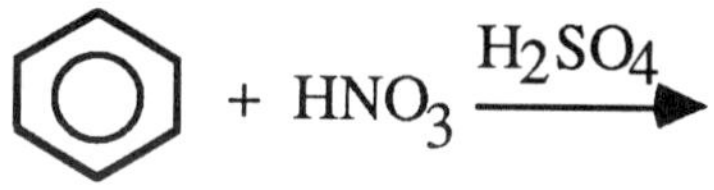


e. $H_3CCH{=}C(CH_3)_2 + H_2O \xrightarrow{H^+}$

f. 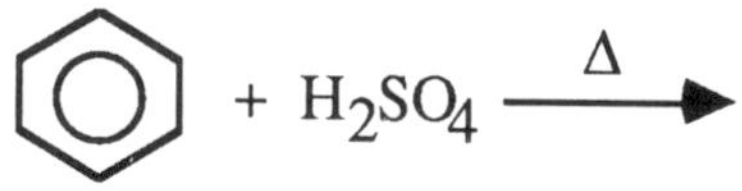


8. Markovnikov's rule states ____________________________________________________
______________________________________________________________________.

9. The structural formula for 6-chloro-2,3,4-trinitrophenol is: ______________________

10. Draw the products for the sequence of the two reactions shown below:

$HC{\equiv}CCH_3$ + $Cl_2$ ——> ______________________
product 1

______________________ + $H_2O$ ——> ______________________
product 1

11. Squalene is a terpene compound involved in the biosynthesis of cholesterol in the body. The condensed structure is shown below and can be regarded as a polymer of isoprene monomer units.

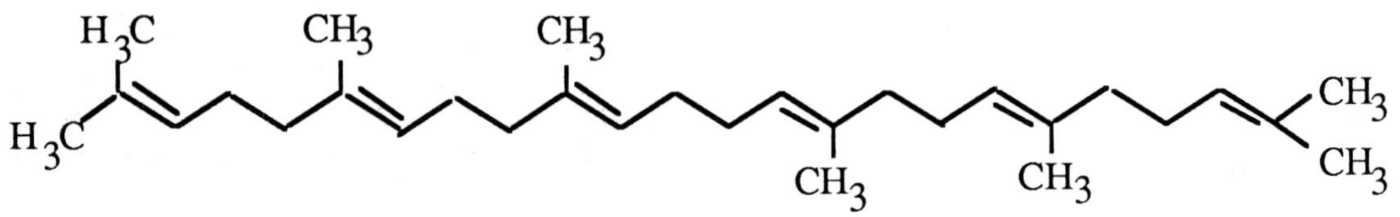


a. The structural formula for an isoprene unit is ______________________.

b. Although the many structures of terpenes are very diverse, they all share two common characteristics.

(i) ______________________

(ii) ______________________

c. Determine the number of isoprene units in squalene and circle these units in the structure shown.

number of isoprene units = ______________________

## ANSWERS TO SELF-TEST QUESTIONS

Multiple Choice

1. a, b, c
2. c
3. b
4. a
5. a, d
6. b, d
7. d
8. a
9. a, b
10. a
11. b
12. c
13. b, d

Completion

1. a. 1,4-cyclohexadiene
   b. 2-bromo-1-butene
   c. trans-1,2-dibromo-1-pentene
   d. benzyl-1-cyclopropene
   e. 1,3-chloronitrobenzene or meta-chloronitrobenzene
   f. 1-bromo-4-methyl-2-pentene

2. Substitution or replacement of a hydrogen on an aromatic ring with the $-SO_3H$ group.

3. a. phenol
   b. acetylene
   c. benzene
   d. pyridine
   e. naphthalene
   f. polyethylene

4. a. 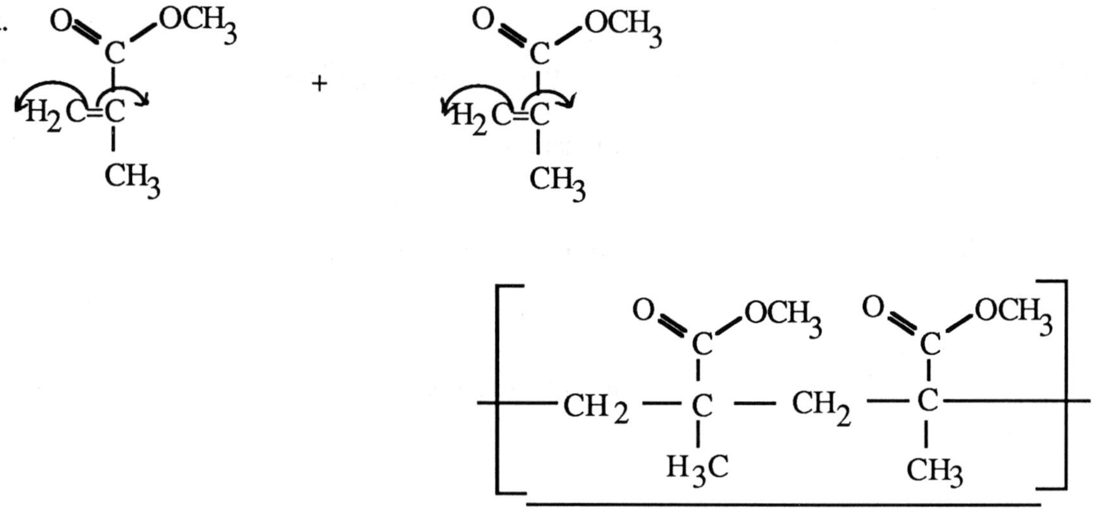

b. 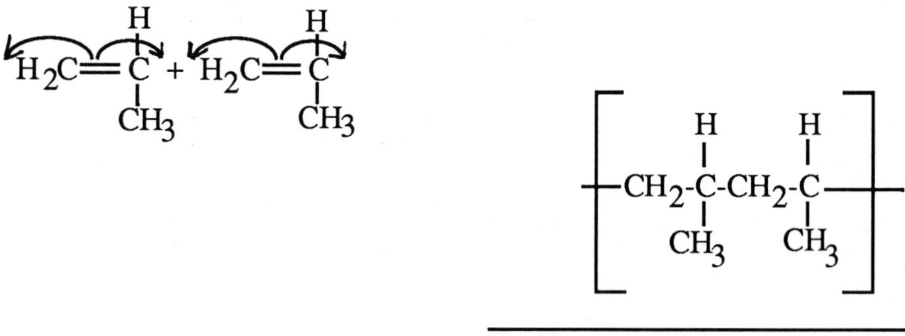

c. addition
d. two different monomer units

5. lower

6. sp, 50, linear, 2
$sp^2$, 33, 67

7. Product | Reaction Type

| | Product | Reaction Type |
|---|---|---|
| a. | $CO_2 + H_2O$ | oxidation |
| b. |  | aromatic substitution reaction |
| c. | 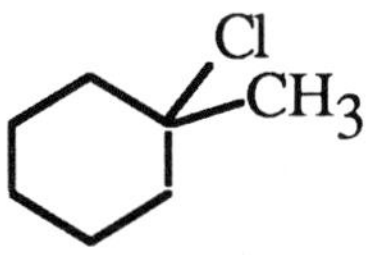 | addition reaction, in accord with Markovnikov's rule |
| d. | 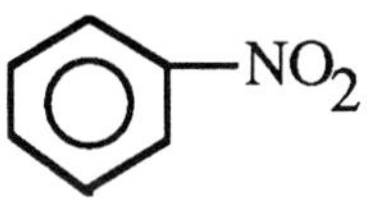 | aromatic substitution reaction (nitration) |
| e. | 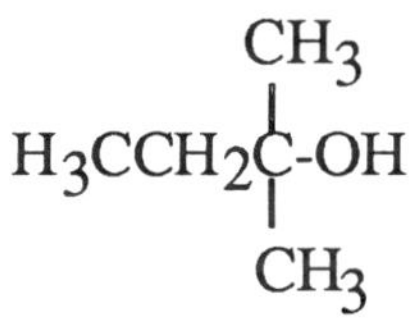 | addition reaction - a hydration in accord with Markovnikov's rule |
| f. | 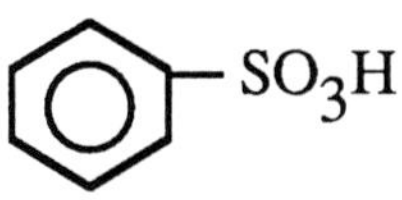 | aromatic substitution (sulfonation) |

8. When an unsymmetrical reagent such as HX adds to an alkene or alkyne, the hydrogen from HX adds to the carbon atom of the double bond which already has the greater number of hydrogen atoms.

9. 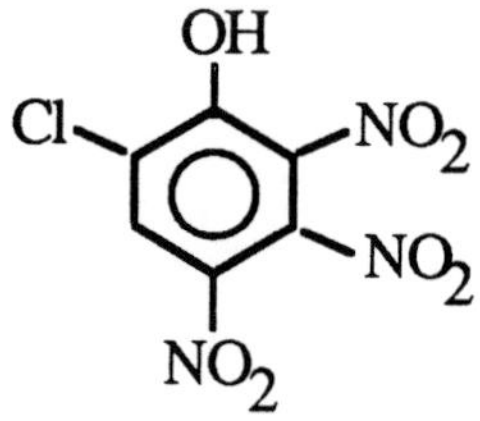

10. 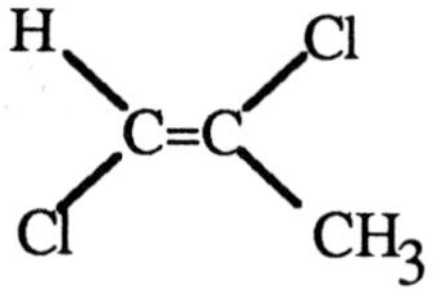 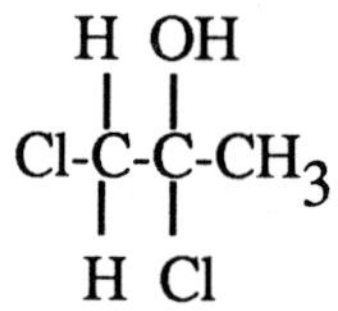

only trans-isomer shown

11. a. 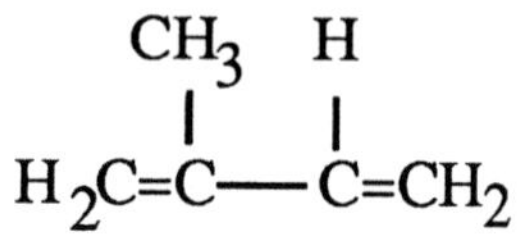

b. (i) all are synthesized in plants
(ii) all follow the isoprene rule

c. 6 isoprene units

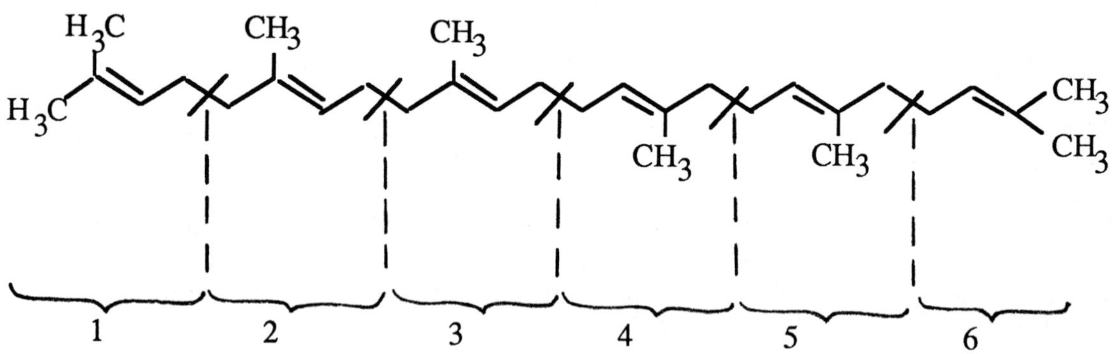

More wine, less wits, you know; wine makes a man sing even if he is a rare scholar, makes him titter and chuckle, aye, makes him dance a jig, and makes him blurt out what were better kept to himself.

- Odyssey, XIV, 464

# 12

# Alcohols, Phenols, Ethers and Halides

## CHAPTER OBJECTIVES

After you have studied the chapter and worked the assigned exercises in the text and the study guide, you should be able to do the following.

1. Identify compounds which are alcohols, phenols, ethers and alkyl- or aryl halides.
2. Given the IUPAC names for simple alcohols, draw the structural formula for each which includes all the atoms.
3. Given the structural formula for an alcohol, name the compound according to the IUPAC rules.
4. Write out the dehydration reactions for alcohols.
5. Draw the structures for primary, secondary and tertiary alcohols.
6. Write out the reactions for the oxidation of primary, secondary and tertiary alcohols and characterize all the possible products formed.
7. Discuss the role of hydrogen bonding in understanding the solubility characteristics of alcohols and ethers.
8. Characterize the similarities and differences in the physical properties of alcohols, ethers and saturated hydrocarbons of comparable molecular weight.
9. Draw the structural formulas for thiols and disulfides and write out chemical reactions associated with each.
10. Characterize alcohols, phenols and thiols according to their acidity.

11. Name the alkyl and aryl halides, draw their structural formulas and point out a major characteristic or use for some of the common ones.

12. Prepare a summary sheet for the reactions of alcohols, thiols and disulfides.

13. Define the important terms and comparisons in this chapter and give specific examples where appropriate.

## IMPORTANT TERMS AND COMPARISONS

Functional Group
Alcohol, Ether and Phenol
Primary, Secondary and Tertiary Alcohol
IUPAC
Diol and Triol
Dehydration
Elimination Reaction
Oxidation Reaction
Carboxylic Acid
Aldehyde and Ketone
Ethanol
Proof and Percent Ethanol
Fermentation
Denatured Alcohol
Grain Alcohol and Wood Alcohol
Alkoxy Group
Glycerol
Nitroglycerine
Tetrahydrocannabinol (THC)
Hydrogen Bonding
Hydrophobic
Thiols, Thioethers and Disulfides
Biodegradable
Polychlorinated Biphenyls (PCBs)
Thyroxine

## SELF-TEST QUESTIONS

MULTIPLE CHOICE. Select the correct answer or answers from the choices listed. In some cases, more than one answer is correct.

1. The chemical oxidation of $H_3CCH_2\overset{OH}{\overset{|}{C}}HCH_3$ could produce:

   a. $H_3CCH_2\overset{O}{\overset{||}{C}}CH_3$

   b. $H_3CCH_2OCH_2CH_3$

   c. $H_3CCH_2\overset{H}{\overset{|}{C}}=O$

   d. $CO_2 + H_2O$

2. The compound which is most capable of hydrogen bonding is:

   a. $H_3COCH_2CH_3$

   b. 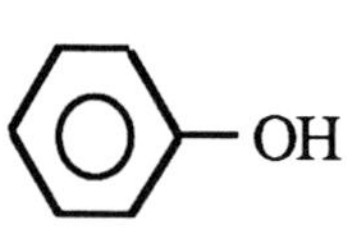

   c. butane

   d. R-S-S-R'

3. Methanol is known as:

a. rubbing alcohol
b. grain alcohol
c. a useful antifreeze
d. wood alcohol

4. Two characteristics of ethers which are potentially dangerous are:

a. solubility in water
b. flammability
c. can be oxidized to peroxides
d. toxicity

5. The compounds which are most known for their unpleasant odors are the:

a. thiols
b. alcohols
c. ethers
d. phenols

6. Tertiary alcohols cannot be reduced because they:

a. are already reduced
b. do not have a H on the C adjacent to the alcohol group
c. are unsaturated
d. readily from a alkene

7. The reaction of thyronine with $I_2$ to produce thyroxine is an example of what type of reaction?

a. addition
b. elimination
c. substitution
d. oxidation

8. The oxidation of RSH results in a:

a. thiol
b. disulfide
c. cyclic thiol
d. useful anesthetic

9. The most acidic compound shown below is:

a. $H_2O$

b. 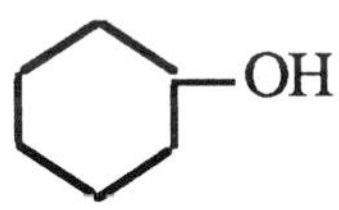

c. 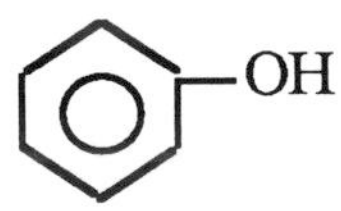

d. ethanol

10. Oxidation of 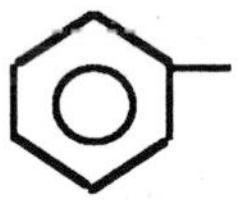 $-CH_2OH$ with $H_2SO_4$ yields:

a. 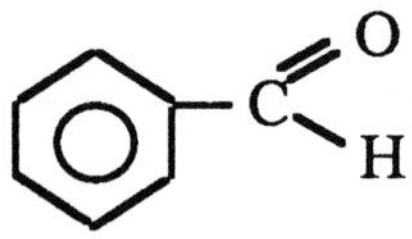

b. 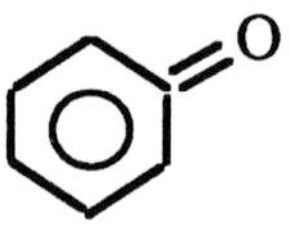

c. 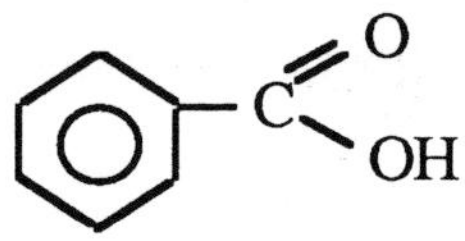

d. 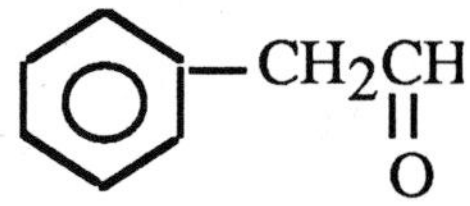

COMPLETION. Write the word, phrase or number in the blank or draw the appropriate structure in answering the question.

1. In the dehydration of an alcohol, two groups are removed from adjacent carbons to produce a double bond. This reaction is referred to as an ____________________.

2. Oxidation of a tertiary alcohol produces ________________, while the oxidation of a primary alcohol yields either an ________________ or an ________________.

3. The relative order of boiling points for a simple alcohol and an ether, both having about the same molecular weight, is:

   ______________ > ______________

4. The relative solubility of the propanol, propane, 1,2-propanediol and ethylmethylether in water is:

   ______________ > ______________ > ______________ > ______________

5. Complete the equations by writing the appropriate structural formula.

a.
```
H3C  CH2OH
 |    |          H2SO4
H3CCHCHCH3      --------->     ____________________
                                  initial product
```

```
                   [O]
____________    ---------->    ____________________
initial product                   final product
```

b. ____________ $\xrightarrow{H_2SO_4}$ $H_3C\text{-}O\text{-}CH_3$

c. ____________ $\xrightarrow{[H]}$
```
     CH3
      |
HSCH2CHCH2SH
```

d. $C_6H_{12}O_6$ (sugar) <u>fermentation enzyme</u> ______________ + ______________

e.  + NaOH -------> ______________ + ______________

f. 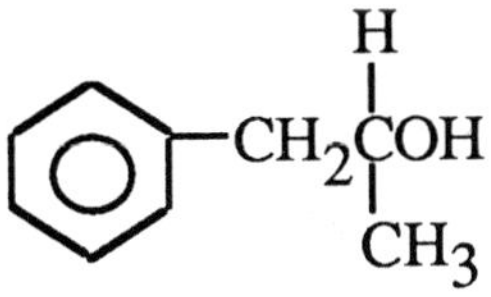 --------> ______________

g. $H_3CCH_2OH$ $\xrightarrow[180^\circ]{H_2SO_4}$ ______________

h. 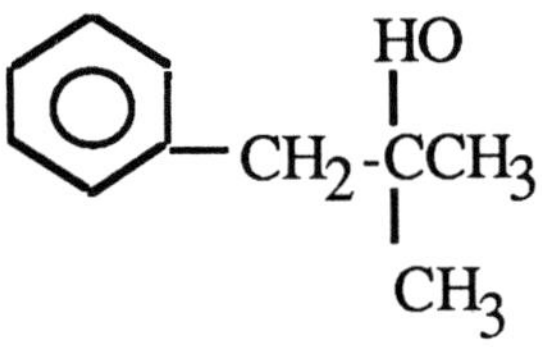 $\xrightarrow[H_2SO_4]{K_2Cr_2O_7}$ ______________

i. 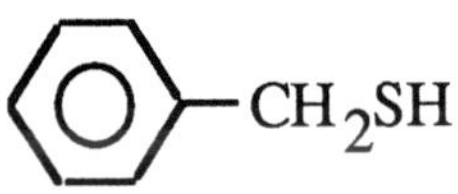 $\xrightarrow{[O]}$ ______________

j. $CH_2{=}CH_2 + H_2O$ $\xrightarrow{H_2SO_4}$ ______________

k. $CH_2CHCH_2 + HNO_3$ $\xrightarrow{H_2SO_4}$ ______________
HO OH OH

6. Characterize each of the following molecules as alcohols, phenols, ethers or none of these:

a. $H_3CCHO$ ______________

b. 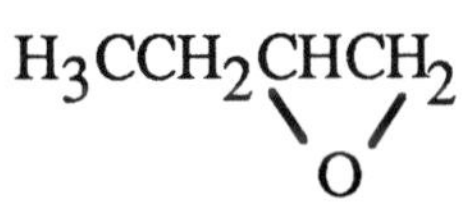 ______________

c. $HOCH_2CH_2OH$ ______________

d. 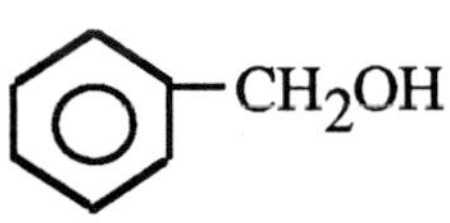 ______________

e. 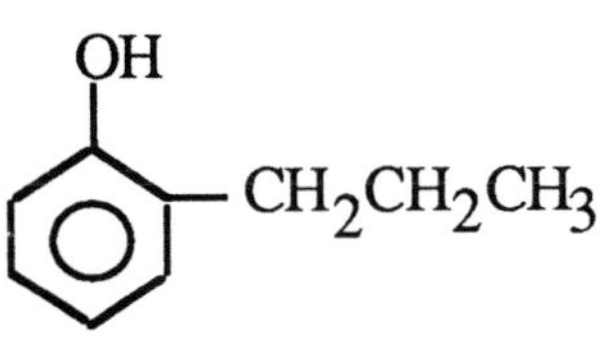 ______________

7. Give suitable names for the following compounds.

a. 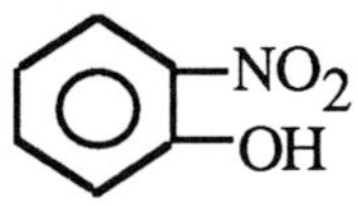

b. 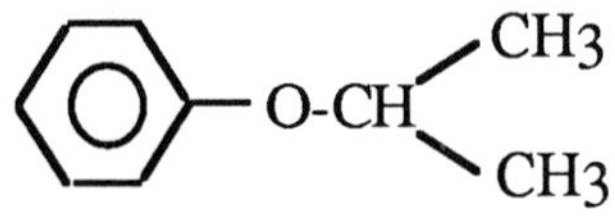

c. 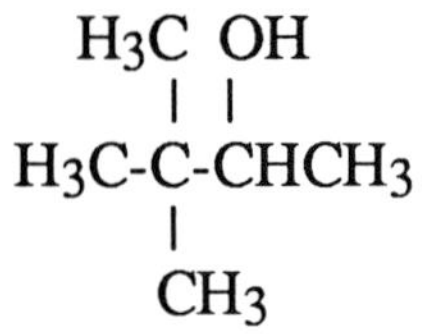

d. 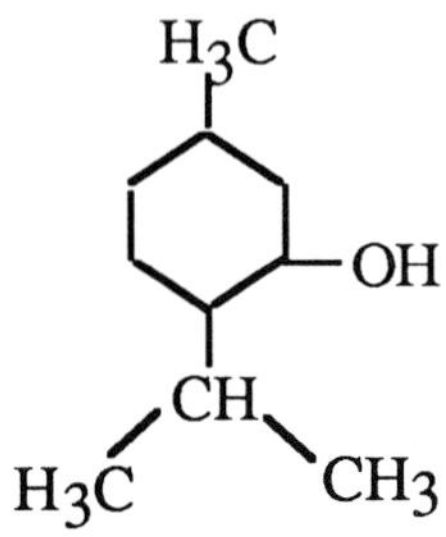

e. $H_3CCH_2CH_2SH$

8. Write out the structural formula for the following compounds:

a. butylmethylether

b. 3-methylphenol

c. 1,2-pentanediol

d. dichlorodifluoromethane

e. DDT or 1,1,1-trichloro-2,2-bis(p-chlorophenyl)ethane

9. Compounds which can be degraded by natural processes in the environment are referred to as ________________.

## ANSWERS TO SELF-TEST QUESTIONS

Multiple Choice

1. a
2. b
3. d
4. b, c
5. a
6. b
7. c
8. b
9. c
10. a and/or c

Completion

1. elimination
2. no reaction; aldehyde, carboxylic acid
3. alcohol > ether
4. 1,2-propanediol > propanol > ethylmethylether > propane
5.

a. 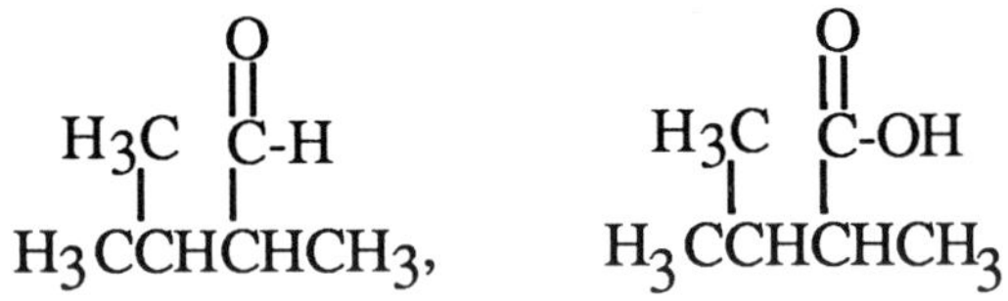

b. $H_3COH$

c. 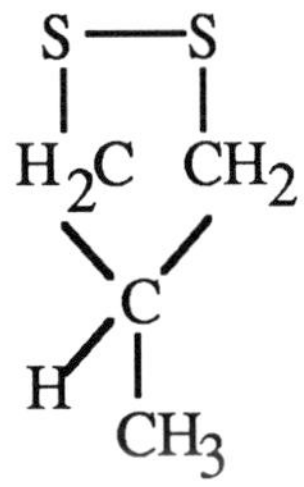

d. $2CH_3CH_2OH + 2CO_2$

e. 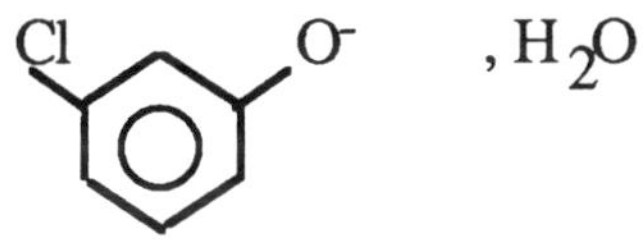

f. $H_2SO_4$, 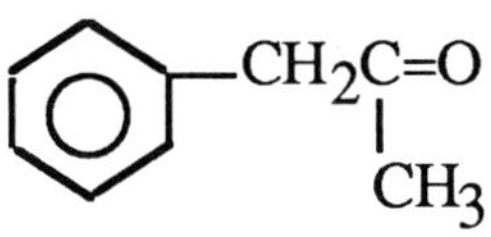

g. $H_2CCH_2$

h. no reaction

i. 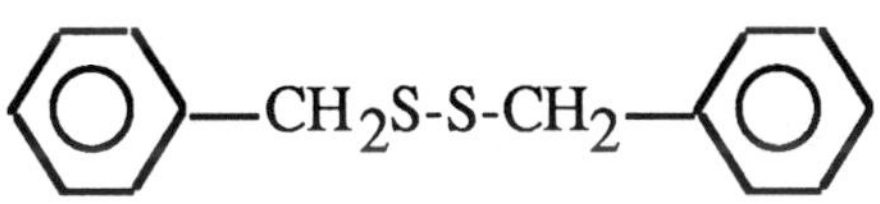

j. $H_3CCH_2OH$

k. 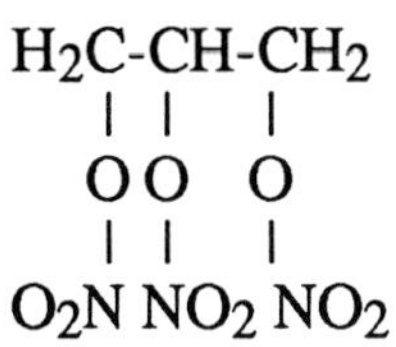

6. a. none of these
   b. ether
   c. alcohol
   d. alcohol
   e. phenol

7. a. 2-nitrophenol
   b. isopropylphenylether
   c. 3,3-dimethyl-2-butanol
   d. menthol or 2-isopropyl-5-methylcyclohexanol
   e. propanethiol

8. a. $H_3C\text{-}O\text{-}CH_2CH_2CH_2CH_3$

   b.

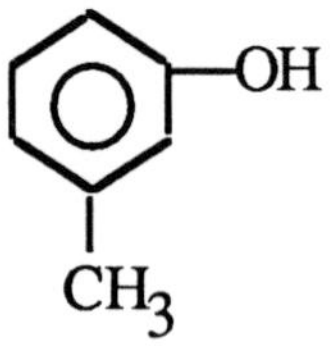

   c. $H_2C\text{-}CHCH_2CH_2CH_3$
   | |
   HO OH

   d.

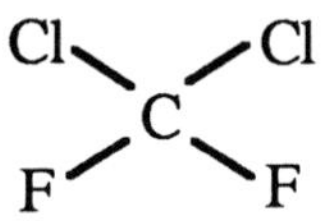

   e.

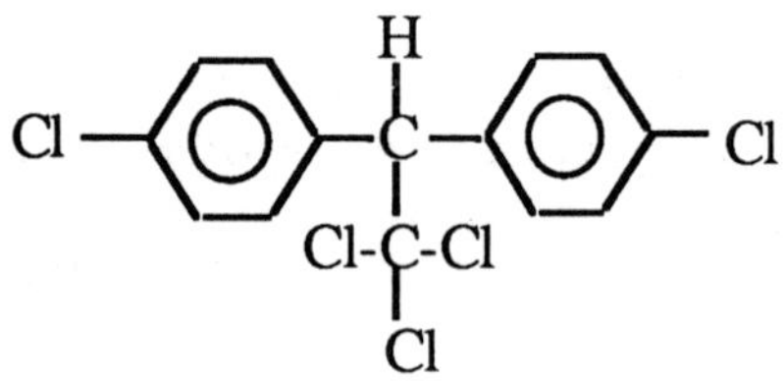

9. biodegradable

Man's mind stretched by a new idea
never goes back to its original dimensions.
- paraphrase from Oliver Wendell Holmes

# 13

# Aldehydes and Ketones

## CHAPTER OBJECTIVES

After you have studied the chapter and worked the assigned exercises in the text and the study guide, you should be able to do the following.

1. Write out the characteristic functional group for an aldehyde and a ketone and contrast this to the structure and bond types found in functional groups presented in previous chapters.

2. Given the structural formula for an aldehyde or ketone, name the compound with the IUPAC nomenclature.

3. Given the IUPAC or common name for an aldehyde or ketone, write out its structural formula.

4. Choosing organic compounds of approximately the same molecular weight, compare the physical properties of an alkane, ketone, aldehyde, alcohol and ether.

5. Write out reactions for the preparation of aldehydes and ketones which have 2, 3 and 4 carbons.

6. Write out the balanced reactions for a positive Tollens and Benedict's test.

7. Describe the basis for using these tests to distinguish between an aldehyde and ketone.

8. Clearly write out the general structure for an aldehyde, a hemiacetal and an acetal. Provide at least one specific example of each.

9. Draw the general structure for a ketone, a hemiketal and a ketal. Provide at least one specific example of each.

10. Write out the chemical reaction for the formation of a hemiacetal and an acetal. Provide examples using at least two different aldehydes.

11. Write the chemical reaction for the formation of a hemiketal and ketal. Provide examples using at least two different ketones.

12. Using a molecule which has both an alcohol and an aldehyde functional group in it, show how a cyclic hemiacetal, and then on further reaction, a cyclic acetal, is formed.

13. Write out the reaction for the formation of a cyclic hemiketal, and also the cyclic ketal which will form after further reaction with additional alcohol.

14. Write out equations for the reduction of aldehydes and ketones to produce alcohols.

15. Draw the structural formulas for formaldehyde, acetaldehyde and acetone and describe two characteristics of each compound.

16. Define the important terms and comparisons in this chapter and give specific examples where appropriate.

## IMPORTANT TERMS AND COMPARISONS

Carbonyl Group
Aldehyde and Ketone
IUPAC
Suffixes -al and -one
Primary and Secondary Alcohol
Tollens and Benedict's Test
Hemiacetal and Acetal
Hemiketal and Ketal
Acetal and Ketal
Cyclic Acetal
Cyclic Ketal
Reduction
Oxidation
NADH
Keto and Enol Forms
Formalin
Bakelite
Thermosetting Plastic
Acetone

In Chapters 10-13, we have increasingly expanded the coverage of organic chemistry - the study of carbon containing compounds with various functional groups (i.e., reactive groups), with an emphasis on the characteristic physical properties and chemical reactions associated with each group. The major functional groups presented to this point are the following.

## MAJOR FUNCTIONAL GROUPS

**I.** Contains Only C and H (Hydrocarbons)

A. Saturated

```
           |  |
Alkanes   -C-C-
           |  |
```

**II.** Contains O, C and H

A. Carbonyl Groups

```
             O
             ||
Aldehyde   R-CH

             O
             ||
Ketone     R-C-R'
```

B. Unsaturated

Alkene 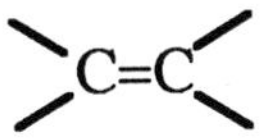

Alkyne -C ≡ C-

B. Others

Alcohol R-OH

Ether R-O-R'

Phenol 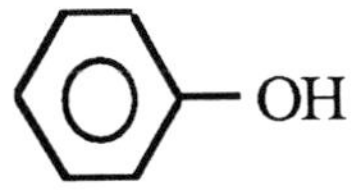

**III.** Contains S, C and H

R-SH Thiol or Mercaptan

R-S-S-R' Disulfide

Organize and review the summary sheets of the reactions which characterize each functional group. The importance of making and continually reviewing (even casually) these sheets throughout the course will become vividly clear when you carry out a major review of the material for each hourly exam and the final.

This chapter has focused on aldehydes and ketones, their physical properties and the chemical reactions associated with each functional group. The production of hemiacetals and acetals from the reaction of an aldehyde with an alcohol is of importance here and in the subsequent chapter on carbohydrate chemistry. The following table summarizes these general structures. You should be able to readily recognize and distinguish between these six compounds before going onto the SELF-TEST.

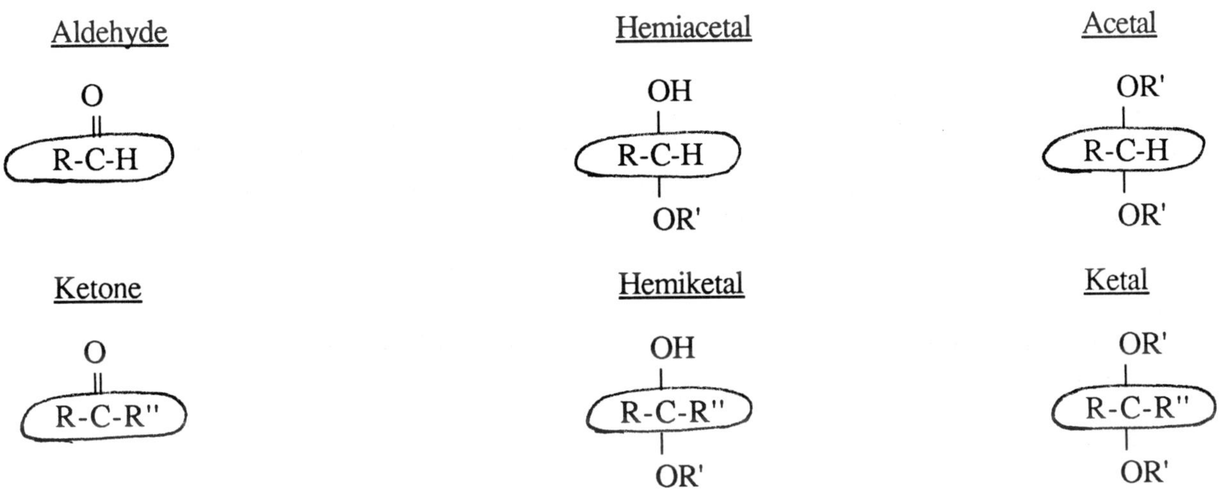


Note that the circled groups remain in the structure for the three related compounds. The R primed group (R') comes from the alcohol group which reacts with the aldehyde or ketone. Write out the reactions for different alcohols, reacting with an aldehyde or ketone, showing the structure of the product. Indicate from which reactant each of the R groups is derived.

## SELF-TEST QUESTIONS

MULTIPLE CHOICE. Select the correct answer or answers from the choices listed. In some cases, more than one answer is correct.

1. Mild oxidation of an aldehyde produces a carboxylic acid. However, mild oxidation of a ketone results in:

   a. an ether
   b. secondary alcohol
   c. carboxylic acid
   d. no reaction

2. Select the compounds which are either aldehydes, hemiacetals or acetals.

   a. 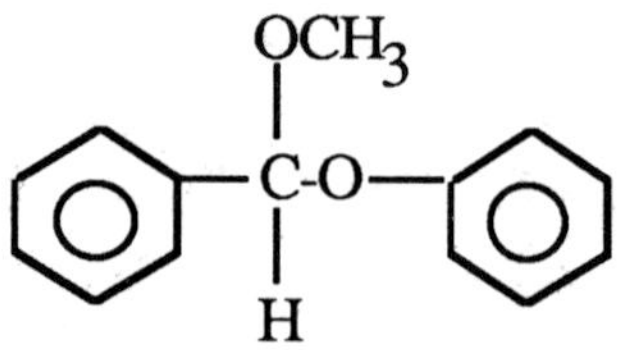

   b. 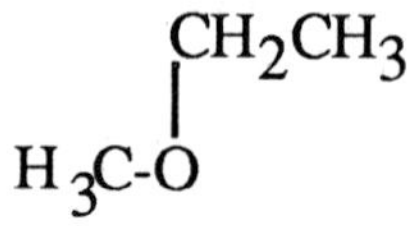

   c. 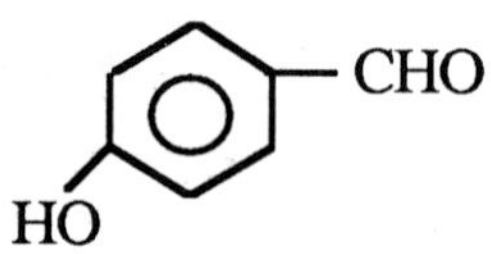

   d. 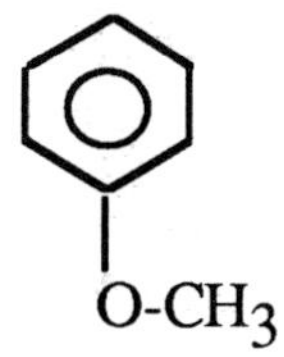

   e. 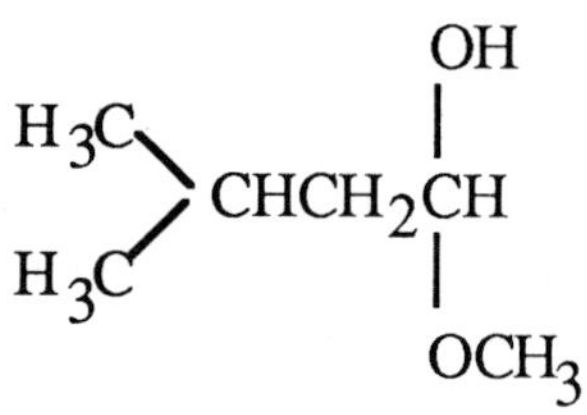

   f. 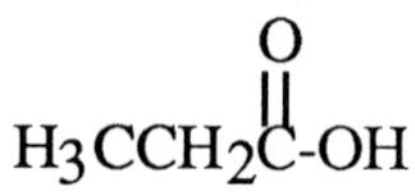

3. Select compounds in question 1 which have a carbonyl group.

4. The reaction of an alcohol with a ketone is an example of a(n):

   a. elimination reaction
   b. addition reaction
   c. hydrolysis reaction
   d. substitution reaction

5. Ketones are prepared by the oxidation of:

   a. secondary alcohol
   b. primary alcohol
   c. tertiary alcohol
   d. all of these

6. The characteristics which occur in a positive Tollens test are:

   a. Oxidation of an aldehyde and reduction of $Ag(NH_3)_2^+$.
   b. Reduction of an aldehyde and oxidation of $Ag(NH_3)_2^+$.
   c. Oxidation of a ketone and reduction to $Cu_2O$.
   d. All of the above.

7. The structural formula for benzaldehyde is:

a. 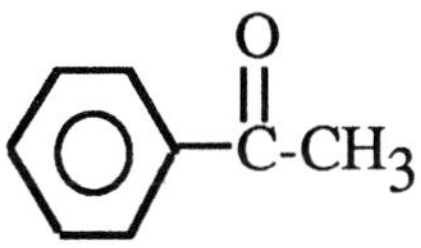

c. 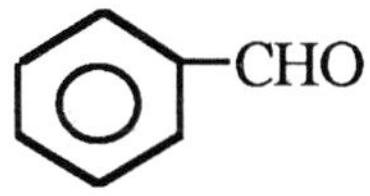

b. 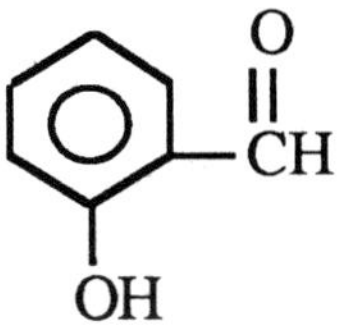

d. 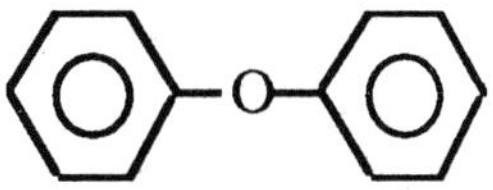

8. Bakelite has the following characteristics:

a. A polymer made from formaldehyde and phenol.
b. A thermosetting plastic.
c. Can be used as an adhesive.
d. All of the above.

9. The structural formula for 2,3-dimethylcyclohexanone is:

a. 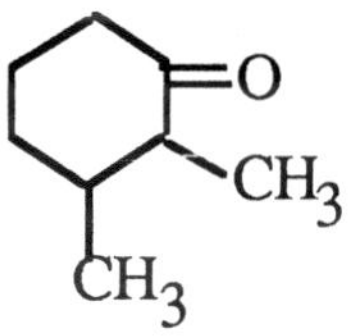

c. 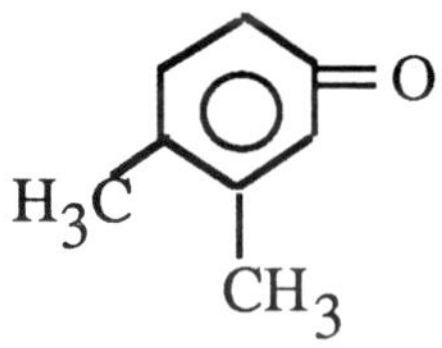

b. 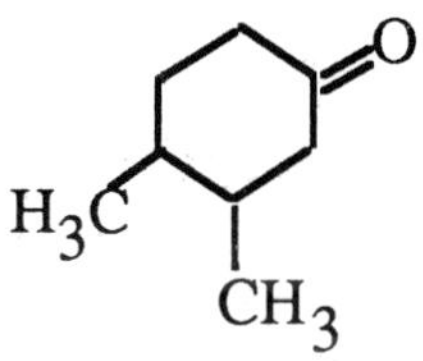

d. 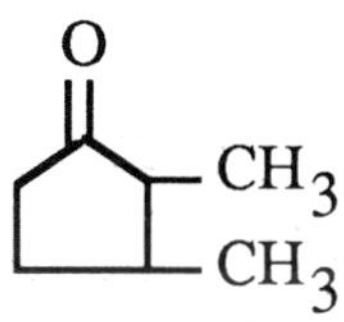

```
                           H3C Br
                            |   |
10. The IUPAC name for H3CCCHCH2CH2CH2CHO
                            |
                            Br
```

a. 5,6-dibromo-6-methylheptanal
b. 1,2-dibromo-1-methylheptanone
c. 5,6-dibromo-1-methylheptanal
d. 1,2-dibromo-1-methylheptanal

11. The structure for one of the forms of guanine is shown below. This form is:

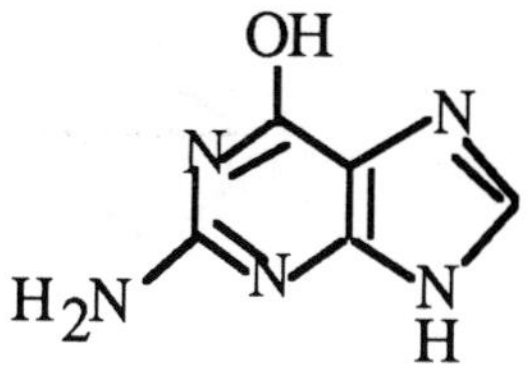


a. a ketone
b. the keto form of guanine
c. the enol form of guanine
d. an acetal

COMPLETION. Write the word, phrase or number in the blank or draw the appropriate structure in answering the question.

1. Write out the IUPAC names for the following molecules.

a. 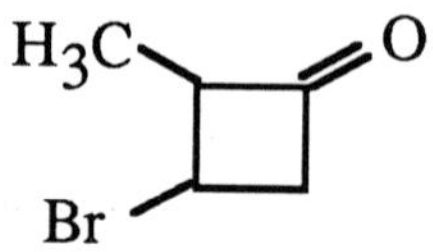

______________________________

b. 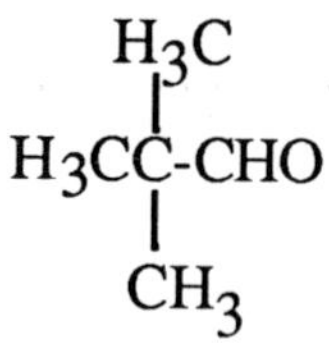

______________________________

c. 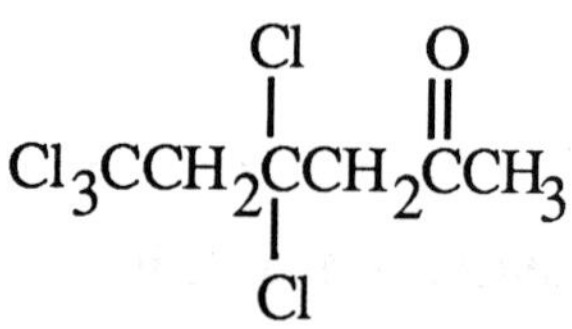

______________________________

d. 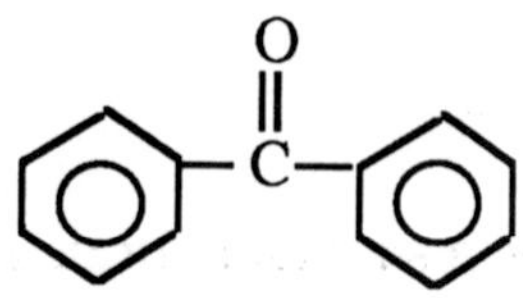

______________________________

2. Given compounds of comparable molecular weight, the order or increasing boiling points for an alcohol, alkane, ether and aldehyde is:

______________ > ______________ > ______________ > ______________

3. Acetone, the most important ______________, can be a product of ______________ metabolism in diabetic patients.

4. Complete the following equations. If no reaction occurs, indicate just that.

a. 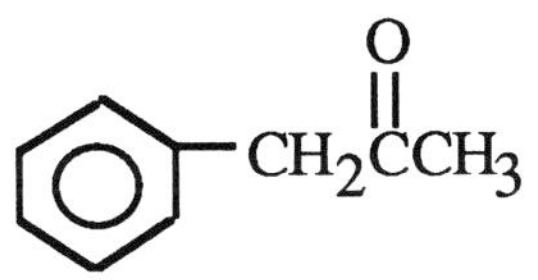 $\xrightarrow{[H]}$ ______________________

b. $H_3CCH_2OH$ $\xrightarrow{[O]}$ ______________________

c. 1 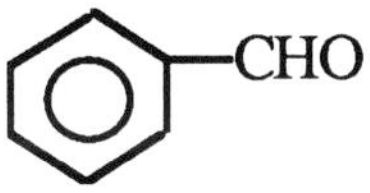 + 1 $CH_3OH$ --------> ______________________

d. $H_3C$–$\underset{|}{C}$HCH$_2$OH (with $H_3C$ on the second carbon) + $H_3C$-C($OCH_2CH_3$)(OH)-$CH_3$ $\overset{H^+}{\rightleftharpoons}$ ______________________

e. $H_3CCH_2\overset{O}{\overset{||}{C}}CH_3$ + $2Cu(OH)_2$ --------> ______________________

f. $\underset{HO}{\underset{|}{C}}H_2CH_2CH_2CH_2CHO$ --------> ______________________

5. The structure of aldehydes would suggest that these compounds ______________ hydrogen bond to each other.

6. The hydrolysis of an acetal yields an ______________ and ______________.

7. A ketal has ______________ ether linkages in it and the general structure is

______________________________________

8. The oxidation of 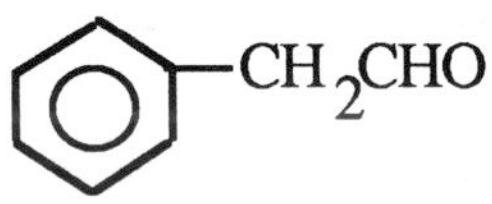 yields the molecule with structural formula

______________________________________

9. A biological reducing agent for ketones and aldehydes in organisms is ______________, which is abbreviated ______________.

10. The mixing of an aldehyde with an acid or base yields ______________ reaction because the aldehyde is ______________ acidic or basic in character.

11. Draw the (a) keto from of the compound, $H_3C\overset{OH}{\overset{|}{C}}=CHCH_3$ and (b) the enol form for the compound

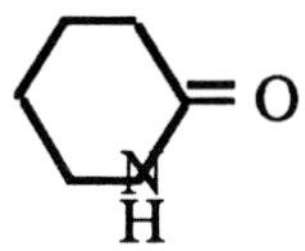

## ANSWERS TO SELF-TEST QUESTIONS

Multiple Choice

1. d
2. a (acetal), c (aldehyde), e (hemiacetal)
3. c, f
4. b
5. a
6. a
7. c
8. d
9. a
10. a
11. c

Completion

1. a. 3-bromo-2-methylcyclobutan-1-one
   b. 2,2-dimethylpropanal
   c. 4,4,6,6,6-pentachlorohexan-2-one
   d. diphenylketone

2. alcohol > aldehyde > ether > alkane

3. ketone, lipid

4. a.

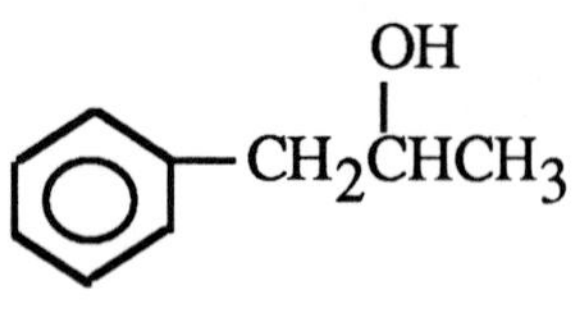

b. $H_3CCHO$

c.

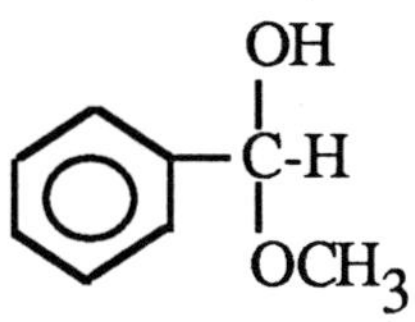

d.

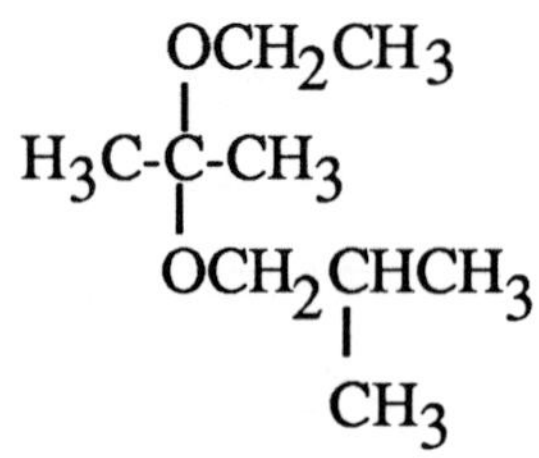

e. no reaction

f.

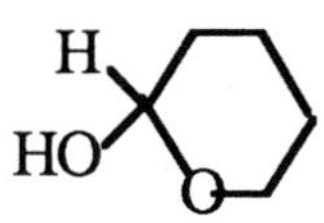

5. do not

6. alcohol and aldehyde

7. two, R-C-R", in which the R' groups are derived from the alcohol

$$\begin{array}{c} OR' \\ | \\ R\text{-}C\text{-}R'' \\ | \\ OR' \end{array}$$

8. 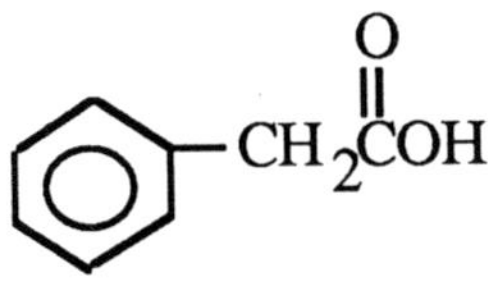

9. nicotinamide adenine dinucleotide, NADH

10. no, neither

11. a. $H_3C\overset{\overset{\large O}{\|}}{C}CH_2CH_3$

b. 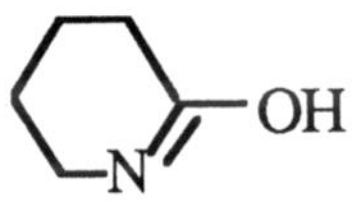

"ASK WHY"
- Sign in the office of Kenneth W. Kirk, former Associate Commissioner, FDA

# 14

# Carboxylic Acids and Esters

## CHAPTER OBJECTIVES

After you have studied the chapter and worked the assigned exercises in the text and the study guide, you should be able to do the following.

1. Recognize the characteristic functional groups in carboxylic acids, carboxylic esters, acyl halides and anhydrides.
2. Write out the structures for these compounds, in addition to those for a carboxylate salt, a dicarboxylic acid, a hydroxyl acid and a phenolic ester. Carefully note the position of the carbonyl and hydroxyl group in all cases.
3. Draw the structure of the fatty acids containing one to six carbons; write out their common names and indicate three characteristics of carboxylic acids.
4. Explain the influence of hydrogen bonding in the physical characteristics of carboxylic acid.
5. Write the reaction for the dissociation of a carboxylic acid and also for the reaction with bases stronger than water.
6. Given the common or IUPAC name for a carboxylic acid or ester, write out its structural formula.
7. Given the condensed structural formula for a carboxylic acid or ester, name the compound by either its common or IUPAC name.
8. Contrast the physical properties of carboxylic acids and corresponding simple esters.
9. Write out the equation for the reaction of a carboxylic acid with:

   a. a base to form a salt of a carboxylic acid
   b. an alcohol to form an ester
   c. another carboxylic acid to form an (carboxylic) anhydride

10. Contrast the reactions and products associated with the acid or base catalyzed hydrolysis of esters.

11. Write out the structures for the anhydrides and esters of phosphoric acid and note the similarities to carboxylic esters.

12. Continue to write out summary sheets for the reactions associated with the new functional groups presented in this chapter. These should include:

    a. Carboxylic acids
    b. Carboxylic esters
    c. Acyl Chlorides
    d. Anhydrides
    e. Phosphoric acid and anhydrides

13. Define the important terms and comparisons in this chapter and give specific examples where appropriate.

## IMPORTANT TERMS AND COMPARISONS

Carboxyl and Carbonyl Group
Carboxylic and Dicarboxylic Acid
Suffix -oic Acid
Carboxylic Acid Dimer
Hydroxy Acid
Ka and pKa
Carboxylic and Phenolic Ester
Catalyst
Flavoring Agent
Acetylsalicylic Acid
Le Chatelier's Principle
Hydrolysis
Condensation Polymer
Dacron
Pyrophosphoric Acid
Alkyl Diphosphates and Triphosphate

## FOCUSED REVIEW

The two basic functional groups in this chapter, from which all other derivatives (compounds) are derived are:

1. Carboxylic acid

```
  O
  ||
R-C-OH
```

2. Phosphoric acid

```
   O
   ||
HO-P-OH
   |
   OH
```

The derivatives, together with the so-called "parent" compound from which they are derived, are shown below.

```
                    O                          O
                    ||                         ||
Acid                R-C-OH                  HO-P-OH
                                               |
                                               OH

                    O          O           O  2-        O  3-
                    ||         ||          ||           ||
Anion(s)            R-C-O-     HO-P-O-     O-P-O        O-P-O
                                  |        |            |
                                  OH       OH           O

                    O          O           O            O
                    ||         ||          ||           ||
Ester(s)            R-C-OR'    HO-P-OR'    R'O-P-OR'    R'O-P-OR'
                                  |            |            |
                                  OH           OH           OR'

                               monoester   diester      triester

                    O  O                   O  O
                    || ||                  || ||
Anhydride           R-C-O-C-R           HO-P-O-P-OH
                                           |  |
                                           OH OH

                                          O  O
Ester of the                              || ||
Anhydride           ______________       -O-P-O-P-OR
                                          |  |
                                          O- O-

                                          O  O  O
                                          || || ||
                                         -O-P-O-P-O-P-OR
                                          |  |  |
                                          O- O- O-

                    O
                    ||
Acyl Chloride       R-C-Cl                 ____________________
```

In each of the derivatives, note exactly how the structure has been changed in going from the parent acid to the derivative compound. The similarities between the carboxylic acid and phosphoric acid should be clearly evident, along with the similarities in the analogous compounds derived from each parent. To become more familiar with these compounds and their chemistries, write out the complete equations for forming each derivative compound from the "parent" acid. Add these reactions to your summary sheets.

SELF-TEST QUESTIONS

MULTIPLE CHOICE. In the following exercises, select the correct answer from the choices listed.

1. The saponification of the carboxylic ester shown below yields the following products:

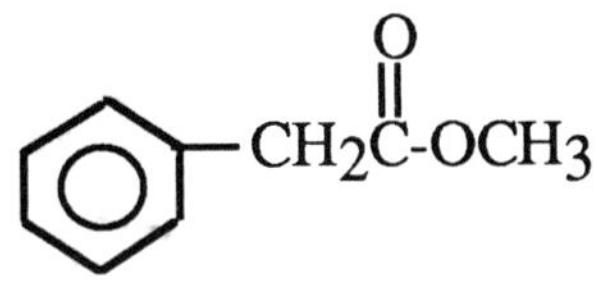

a. 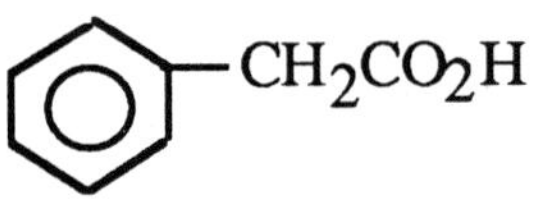 , $CH_3OH$

b. $H_3CCO_2H$, 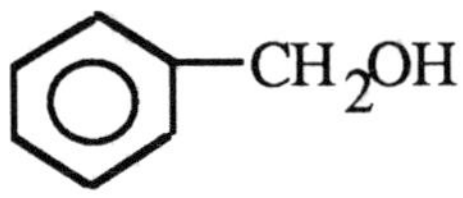

c. 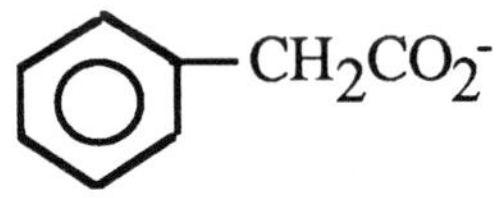 , $CH_3OH$

d. none of the above

2. Carboxylate ion can exist only in water at:

a. pH 7

b. low pH

c. high pH

d. $[H^+]$ greater than $10^{-2}$ M

3. The name of the compound, $H_3CCH_2CH_2CH_2\overset{\overset{O}{\|}}{C}OCH_2CH_2(CH_3)_2$, is

a. isobutyl valerate

b. butyl propionate

c. valeryl isobutyrate

d. isobutyl butyrate

4. Three organic acids are listed below with their pKa values.

| | | pKa value |
|---|---|---|
| Benzoic acid | 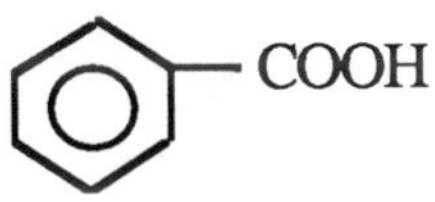 | 4.19 |
| Chloroacetic acid | $ClCH_2COOH$ | 2.86 |
| Formic acid | HCOOH | 3.74 |

The order of increasing acid strength is:

a. chloroacetic acid > formic acid > benzoic acid

b. benzoic acid > formic acid > chloroacetic acid

c. all are the same

d. not determinable from available data

5. The name of the compound, 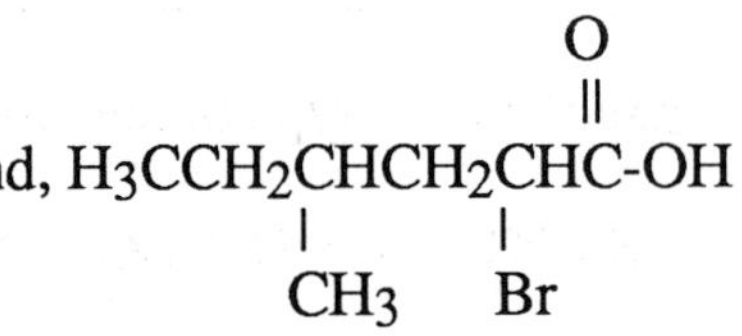 , is

a. 1-bromo-3-methylhexanoic acid
b. 2-bromo-4-methylhexanoic acid
c. 2-bromo-4-methylpentanoic acid
d. 5-bromo-3-methylpentanoic acid

6. The hydrolysis of what ester produces butyric acid and phenol.

a. phenyl butyrate
b. benzyl propionate
c. butyl benzoate
d. phenyl propionate

<u>COMPLETION</u>. Write the word, phrase or number in the blank or draw the appropriate structure in answering the questions.

1. Write out the functional group or groups contained in each of the following condensed structural formulas. Some of the functional groups were presented in previous chapters.

a. 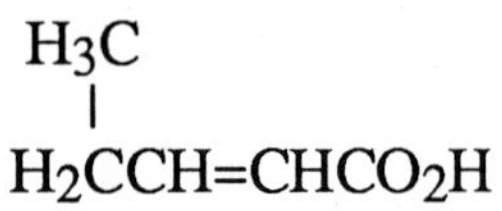 ____________________

b. 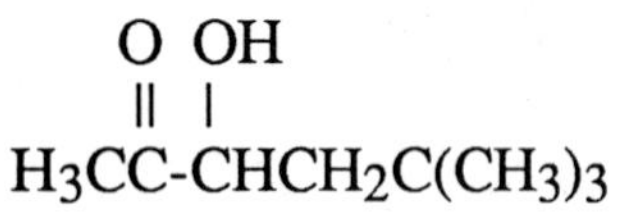 ____________________

c. 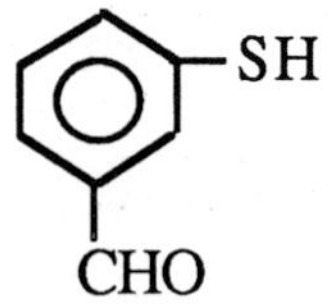 ____________________

d. 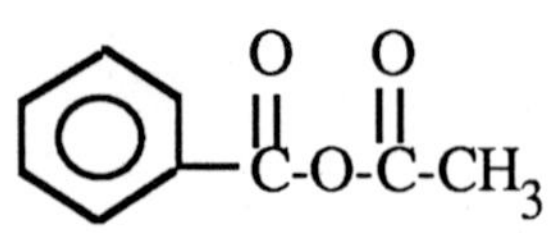 ____________________

e. 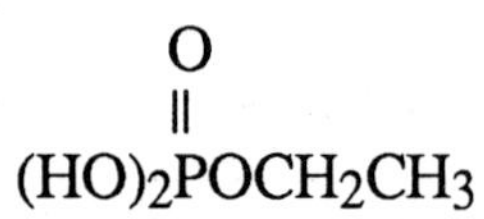 ____________________

f. 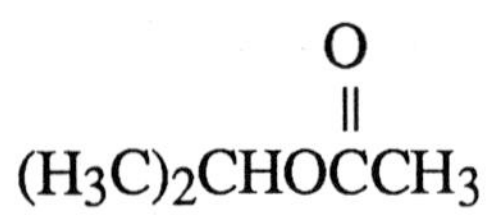 ____________________

g. 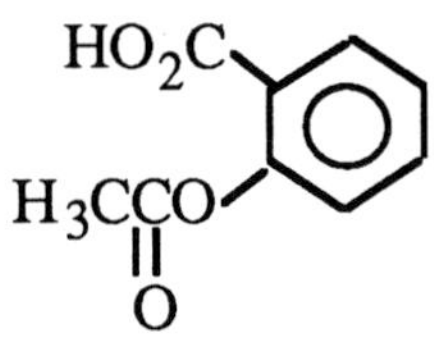 ____________________

h. $H_3CCO_2^-$ ____________________

2. Complete the following equations and name the type of reaction.

a. ____________________ + NaOH -----> $H_3CCO_2^-\ Na^+ + H_2O$

b. 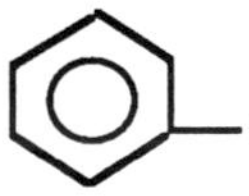$-CO_2H + CH_3CH_2OH \xrightarrow{H^+}$ ____________________

c. ____________________ $\xrightarrow{[O]}$ $(H_3C)_2CHCH_2CO_2H$

d. $H_3CCH_2CO_2H$ + __________ -----> $H_3C\overset{O}{\overset{\|}{C}}\text{-}O\text{-}\overset{O}{\overset{\|}{C}}CH_2CH_3$

e. 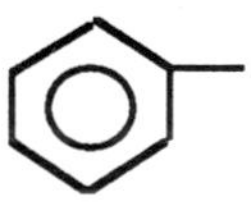$-CO_2H \rightleftharpoons H^+ +$ ____________________

f. __________ + $H_3CCH_2OH$ -----> $\underset{\underset{O}{\|}}{C}\text{-}OCH_2CH_3 + HCl$

3. The structure of the dimer formed by hydrogen bonding between 2 carboxylic acid molecules, $RCO_2H$, is:

4. Polyesters are examples of condensation polymers. To prepare these large molecules, each reactant molecule must have at least two functional groups. Draw out at least one repeating unit produced in the polyester formed in the reaction shown.

$HOCH_2OH$ + $HO_2CCH_2CH_2CH_2CO_2H$ ----->

5. The order of increasing boiling points for an alkane, a carboxylic acid, an ester and an alcohol, all of comparable molecular weight, is:

____ > ____ > ____ > ____

6. The primary reason for the order observed in the compounds in question 5 is ________

______________________________________________.

7. The molecular formula for n-octyl acetate is $H_3C\overset{\overset{\displaystyle O}{\|}}{C}O(CH_2)_7CH_3$. This molecule gives the fragrance of an orange. Draw the structure of the alcohol and the acid from which the ester was made.

____________________ ____________________

alcohol acid

8. Name the following compounds.

a. 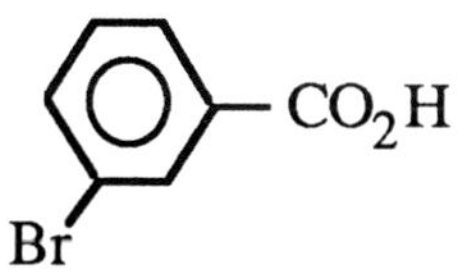 ____________________

b. 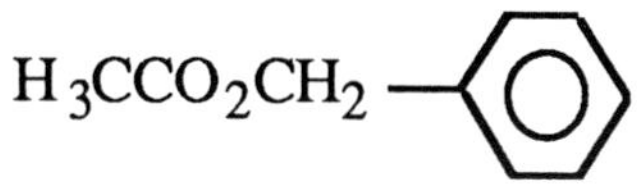 ____________________

c. 

Cl
|
$H_3CCH_2C\text{-}CO_2^-\ K^+$ ____________________
|
Cl

9. The process of phosphorylation involves the transfer of a ______________ from a ______________ to an ______________.

10. The high energy bond in an ester of a pyrophosphoric acid is at the ______________ linkage.

## ANSWERS TO SELF-TEST QUESTIONS

Multiple Choice

1. c
2. c
3. a
4. a
5. b
6. a

Completion

1. a. alkene, carboxylic acid
   b. ketone, alcohol (note that this is not a carboxylic acid)
   c. thiol, aldehyde
   d. anhydride
   e. phosphate (mono) ester
   f. ester
   g. acid, ester
   h. carboxylate anion

2. a. $H_3CCO_2H$; neutralization of (a carboxylic acid)

   b. 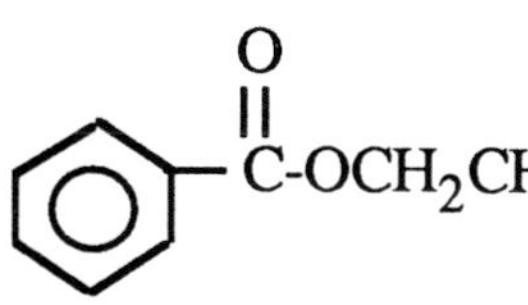; preparation of a (carboxylic) ester

   c. $(H_3C)_2CHCH_2OH$; preparation of a carboxylic acid by oxidation of a primary alcohol.

d. $H_3CCO_2H$; preparation of an anhydride

e. 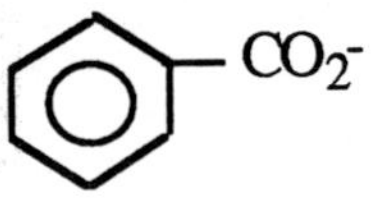 ; reversible dissociation of a carboxylic acid (a weak acid)

f. 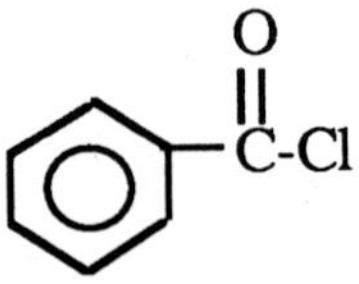 ; reaction of an acyl chloride with an alcohol to produce an ester and HCl

3.
```
    O.....HO-C-R
    ||       ||
R-C-OH.....O
```

4.
```
 O          O               O
 ||         ||              ||
-CC[O-CH2-O-CCH2CH2CH2C-O]CH2O-
```

5. carboxylic acid > alcohol > ester > alkane

6. hydrogen bonding between molecules. The greater the hydrogen bonding between molecules, the more energy it requires to vaporize the liquid and therefore the higher is the boiling point.

7. $H_3C(CH_2)_7OH$; $H_3CCO_2H$

8. a. 3-bromobenzoic acid
   b. benzyl acetate
   c. potassium 2,2-dichlorobutanoate

9. phosphate group, phosphoric anhydride, alcohol

10. P-O-P

Common sense is the collection of
prejudices acquired by age eighteen"
- Albert Einstein

# 15

# Amines and Amides

## CHAPTER OBJECTIVES

After you have studied the chapter and worked the assigned exercises in the text and the study guide, you should be able to do the following.

1. Write out a general structural formula for a primary, secondary and tertiary amine and for a quaternary ammonium salt.
2. Write out the names for simple amines, including those cyclic compounds in which it is necessary to name the $-NH_2$ group.
3. Rationalize the trend in the boiling points for simple primary amines.
4. Describe the hydrogen bonding which takes place (a) between amines and (b) between amines and water.
5. Discuss the influence of hydrogen bonding on the physical properties of amines.
6. Give the names and structures of synthetic amines which are used medicinally.
7. Write out the chemical equations for the reaction of amines with inorganic- and carboxylic acids.
8. Write out the reactions for a primary, secondary and tertiary amine with nitrous acid.
9. Discuss why the use of nitrites as food preservatives is regarded as a potential health hazard.
10. Define the composition of and the general structural formula for an amide.
11. Name a number of substituted amides which are derived from two or three carbon carboxylic acids and simple amines.
12. Name specific amides which are widely used as either (a) pain relievers, (b) drugs or (c) tranquilizers.

13. List the names and structural formulas for the types of organic compounds characterized in Chapters 10-15 and classify them as either acidic, basic or neutral.

14. Write out a general and then a specific equation for the preparation of amides using (a) DDC, (b) acyl chloride and (3) an anhydride as one of the reagents.

15. Write out the equations for the acid and base catalyzed hydrolysis of amides.

16. Name at least a half dozen different alkaloids and indicate their physiological effect on the body.

17. Review the summary of reactions for amines and amides in the text and write out specific reactions for each type.

18. Define the important terms and comparisons in this chapter and give specific examples where appropriate.

IMPORTANT TERMS AND COMPARISONS

Amine and Amide
Primary, Secondary and Tertiary Amine
Hydrogen Bonding
Amphetamine
Bronsted-Lowry Base
Dehydrating Agent
Benzodiazapines
Nylon 66
Barbiturate and Amphetamine
Hydrolysis
Heterocyclic

SELF-TEST QUESTIONS

MULTIPLE CHOICE. Select the correct answer from the choices listed. In some cases, more than one choice will be correct.

1. The name of the compound given below is:

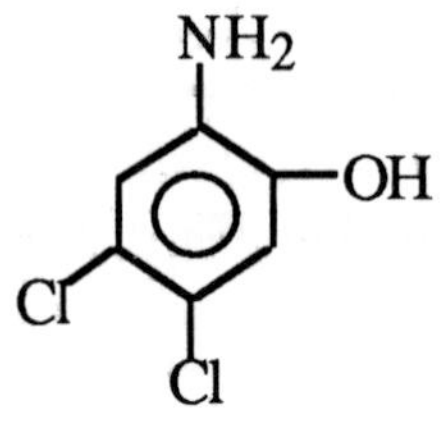


a. 2-amino-4,5-dichlorophenol
b. dichlorohydroxylaniline
c. 2-amino-4,5-dichlorocyclohexanol
d. 3,4-dichloro-1-hydroxylaniline

2. The structure of N,N dibromoaniline is:

a. 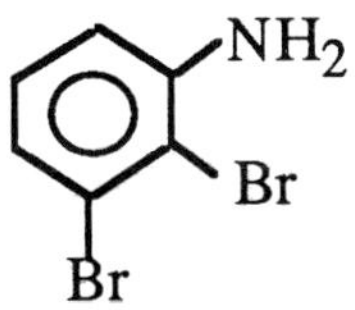

c. 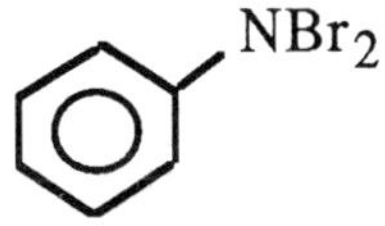

b. 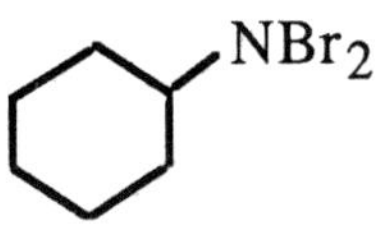

d. 

3. A reaction which is used to distinguish or test for a primary amine involves the reaction of the amine with:

a. HCl
b. HONO
c. DDC
d. $H_3CCH_2Cl$

4. The reaction of nitrous acid with a secondary amine produces:

a. $N_2$ and other products
b. N-nitroso amine plus water
c. N-nitroso amine only
d. generally no reaction

5. Cocaine, atropine and LSD are all examples of:

a. heterocyclic ring compounds
b. alkaloids
c. amides
d. synthetic pain killers

<u>COMPLETION</u>. Answer or complete the questions as indicated.

1. Indicate whether the compounds shown below are amines, amides, both or neither. If the compound contains an amine, indicate whether it is a primary, secondary or tertiary amine.

a. 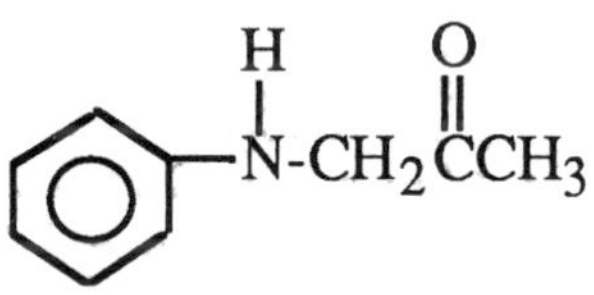 ______________________________

b. 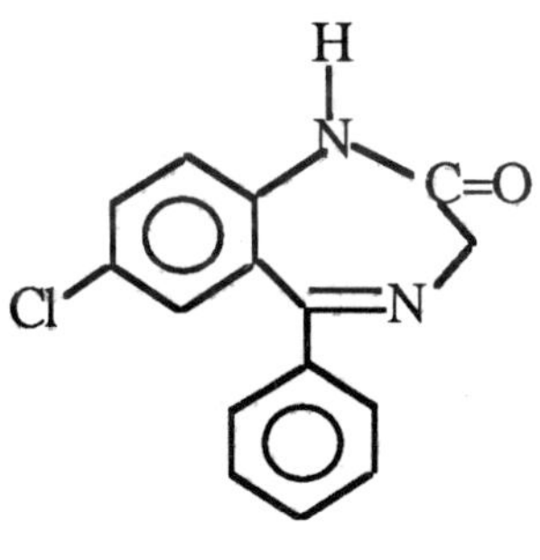

Valium ______________________________

c. Nylon 66 ______________________________

d. 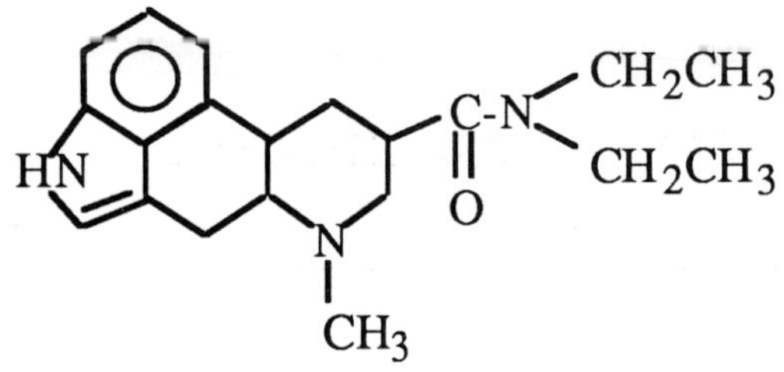

LSD ______________________________

e. 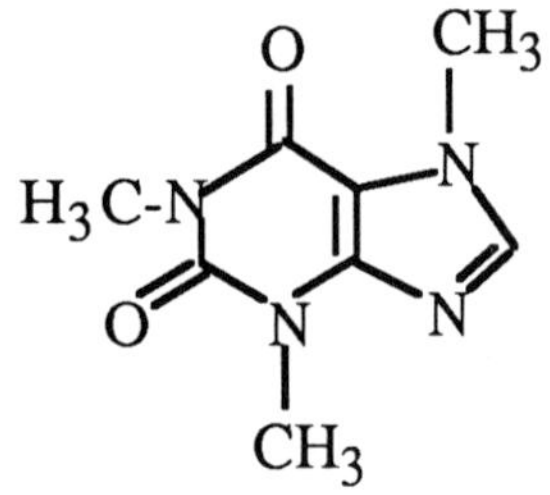

Caffeine ______________________________

2. Name the following compounds.

a. 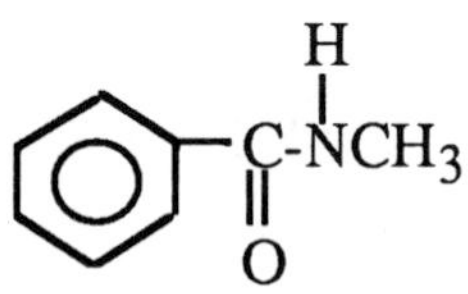 ______________________________

b. 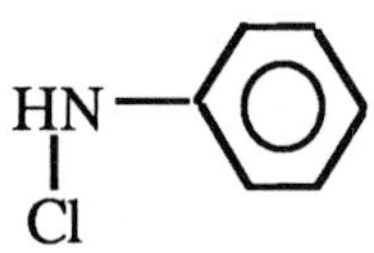 ______________________________

c. 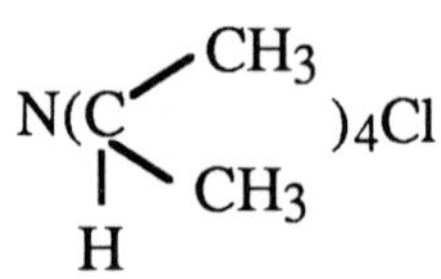 ______________________________

d. $H_2N\overset{\overset{\displaystyle O}{\|}}{C}NH_2$ ______________________________

3. The order of increasing boiling points for the following series of compounds is:

a. $NH_3$, $NH_2CH_3$, $N(CH_3)_3$

______________ < ______________ < ______________

b. $H_3CCH_2CH_2CH_2OH$; $H_2CCH_2CH_2\overset{\overset{O}{||}}{C}CH_3$; $H_3CCH_2CH_2CH_2NH_2$

______________ < ______________ < ______________

4. Complete the following reactions. If no reaction occurs, indicate no reaction.

a. 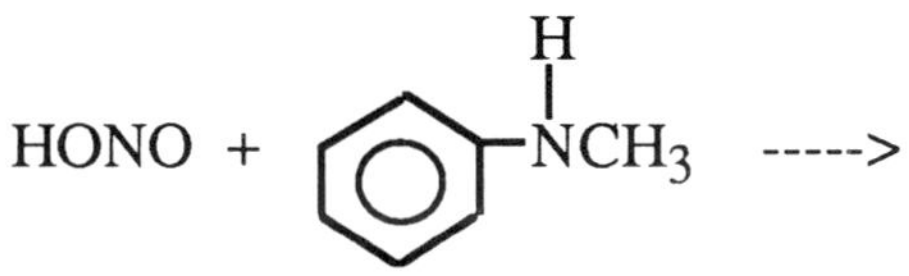

b. 

c. $H_3CCH_2CO_2H + (CH_3)_3N \text{ -----> }$

d. 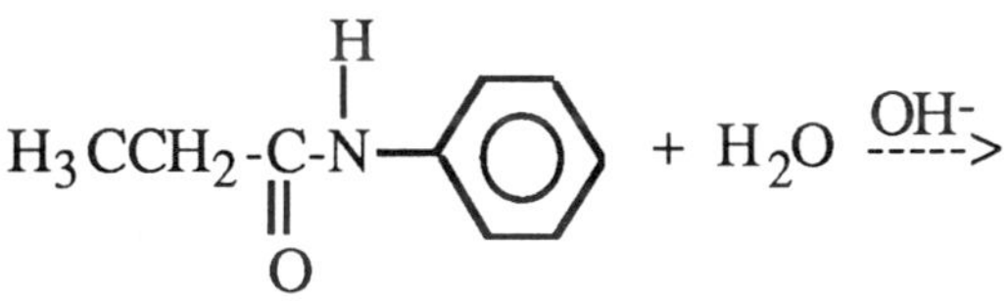

e. $N(CH_2CH_3)_3 + HCl \text{ -----> }$

f. 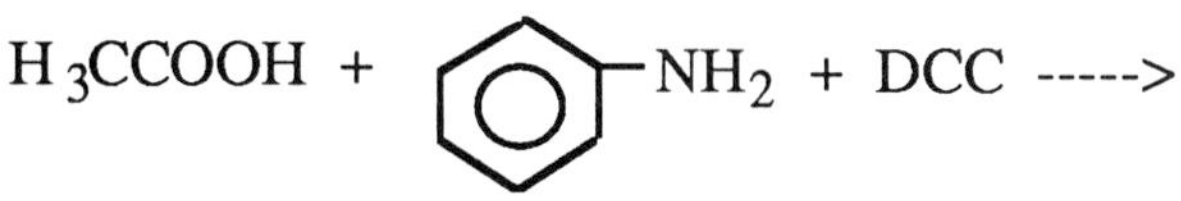

g. $H_2O + H_3C\overset{\overset{CH_3}{|}}{C}H\underset{\underset{O}{||}}{C}NH_2 \xrightarrow{H^+}$

h. $(H_3C)_3N + HONO \text{ -----> }$

5. a. Secondary amines have ______________ hydrogen(s) bound directly to the nitrogen.

b. The compound $[(CH_3CH_2)_4N]^+Br^-$ is a ______________ and can be made by the reaction of ______________ with ____________________.

MATCHING. Match the term in the right-hand column with the most appropriate answer in the left column.

| | | | |
|---|---|---|---|
| 1. | Putrescine | a. | Aspirin substitute |
| 2. | Amphetamine | b. | Downers or sedatives |
| 3. | Nylon 66 | c. | Basic compound extracted from plants |
| 4. | Barbiturates | d. | Pep pills, action similar to that of epinephrine |
| 5. | Librium or Valium | e. | Decomposition product of protein |
| 6. | Hallucinogen | f. | Product of reaction of secondary amine and nitrous acid |
| 7. | Alkaloid | g. | Lysergic acid diethylamide |
| 8. | Acetaminophen | h. | Tranquilizer |
| 9. | N-nitrosoamine | i. | Polyamide product of reaction of hexamethyene diamine and adipic acid |

## ANSWERS TO SELF-TEST QUESTIONS

### Multiple Choice

| | |
|---|---|
| 1. a | 4. b |
| 2. c | 5. a, b |
| 3. b | |

### Completion

1. a. secondary amine
   b. amide
   c. polyamide
   d. secondary and tertiary amines, amide
   e. amides, tertiary amine

2. a. N-methylbenzamide
   b. N-chloroaniline
   c. tetraisopropylammonium chloride
   d. urea

3. a. $N(CH_3)_3 < NH_2CH_3 < NH_3$
   b. ketone < amine < alcohol

4.

a. 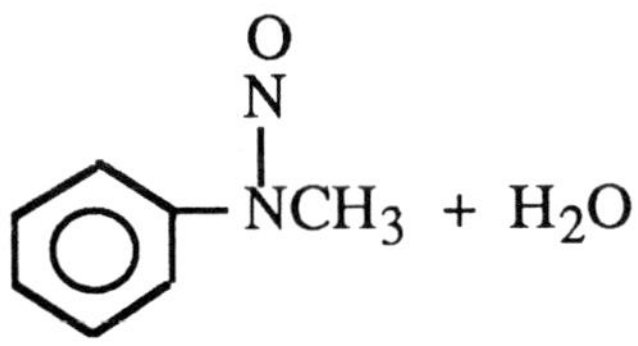

b. no reaction

c. $H_3CCH_2CO_2^- + HN(CH_3)_3^+$

d. 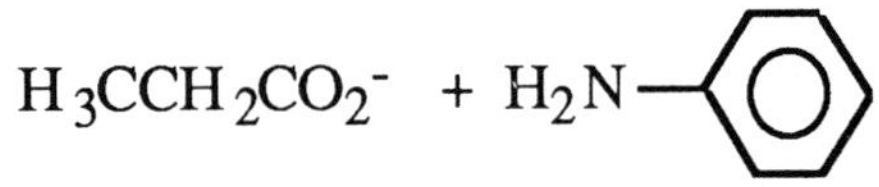

e. $HN(CH_2CH_3)_3^+ \; Cl^-$

f. 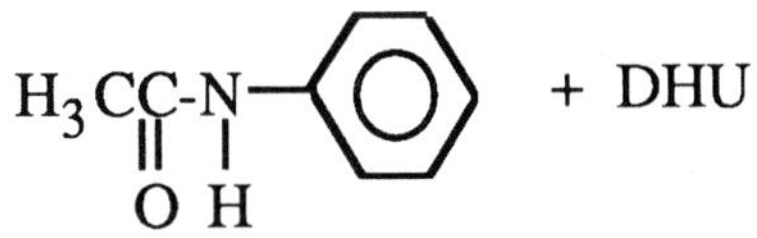

g. $H_3C\text{-}\overset{\displaystyle CH_3 \atop |}{C}HCO_2H + NH_4^+$

h. no reaction

5. a. one

b. quaternary ammonium salt, triethylamine, ethyl bromide

Matching

1. e
2. d
3. i
4. b
5. h
6. g
7. c
8. a
9. f

Burnt Sugar Icing Recipe
2/3 cup granulated sugar
1/3 cup boiling water
3 cups powdered sugar
2 tablespoons butter
- H. Morgan
You Can't Eat That!

# 16

# Carbohydrates

## CHAPTER OBJECTIVES

After you have studied the chapter and worked the assigned exercises in the text and the study guide, you should be able to do the following.

1. Define carbohydrate and indicate a number of specific uses of carbohydrates by humans and plants.

2. Draw the D-form of the aldoses and ketoses which have three, four and five carbon atoms. Identify the chiral atoms and the position which defines the molecule as being either the D- or the L-form.

3. Calculate the maximum number of stereoisomers for aldoses and ketoses with 3, 4 and 5 carbon atoms. Draw the Fischer projection formulas for these and identify the enantiomeric pairs of molecules, and the diastereomers. Specify the D- and L-form of these sugars.

4. Indicate the physical and chemical properties which differ in enantiomers.

5. List four properties which are identical for enantiomers.

6. Calculate the specific rotation of a substance from the observed optical rotation, its concentration and the path length of the cell which the light passes through.

7. Calculate the concentration of an optically active species in solution from the specific rotation, the observed optical rotation and the path length of the light passing through it.

8. Clearly distinguish the difference between a structural and a stereoisomer. Give an example of each type.

9. Describe in words the solution equilibrium for a monosaccharide such as β-D-glucose. Also draw the open chair and ring structures for the molecules in the equilibrium. Identify the hemiacetal units in the Haworth formulas.

10. Write out the reactions for D-glucose undergoing (a) reduction, (b) oxidation and (c) phosphorylation. Note the position at which the reaction takes place.

11. List three monosaccharides and three disaccharides of importance. Name the products obtained from the hydrolysis of the specific disaccharides.

12. Draw the Haworth structure for α-maltose, pointing out (a) the acetal unit, (b) the reducing ring and (c) the glycosidic linkage.

13. Describe the structural difference between amylopectin and amylose. Draw a minimum unit of this structure to show the difference.

14. Describe the major structural difference between starch and cellulose; explain the significance of this with regard to human metabolism.

15. Characterize the linkage types found in the hyaluronic acid and heparin; explain why both polysaccharides are considered acidic.

16. Define the important terms and comparisons in this chapter and give specific examples where appropriate.

## IMPORTANT TERMS AND COMPARISONS

Chiral or Asymmetric Carbon
Chiral Molecule
Enantiomer and Diastereomer
Nonsuperimposable Mirror Image
Plane Polarized Light
Dextro- and Levorotary
Optically Active Molecule
Racemate and Racemic Mixture
Fischer Projection and Haworth Formula
$2^n$
Mono-, Di-, Oligo- and Polysaccharides
Glucose, Dextrose, Grape Sugar or Blood Sugar
Hemiacetal and Acetal
Hemiketal and Ketal
Pyranose and Furanose Rings
Anomer and Anomeric Carbon Atom
α- and β-D-Glucose
D- and L-Glucose
Aldose and Ketose
Invert Sugar
Saccharin
Reducing and Non-Reducing Sugar
Glycoside and Glucoside
Aldose Reductase
Aldonic and Uronic Acids
Sorbitol
Blood Groups
Antigen and Antibody
"Locked" Ring of a Disaccharide
Sucrose, Lactose and Maltose
Starch and Cellulose
Amylose, Amylopectin and Glycogen
Hydrolysis
α(1--->4) and β(1--->6) Glycosidic Linkages
Diabetes Mellitus
Cellulase
Hypo- and Hyperglycemia
Acidic Polysaccharide
Collagen
Hyaluronic Acid and Heparin

FOCUSED REVIEW

As we embark now into the study of carbohydrates and begin to focus on the major areas of **biochemistry**, it will become increasingly clear that many of the chemical concepts presented in previous chapters provide the basis for an understanding of biochemistry. As you encounter these basic concepts, do not be shy about returning to the previous chapter to review and reclaim a clear understanding of any vague terms. For example, a few of the concepts and reactions which were presented in previous chapters and are relevant to this chapter are presented below. You may want to extend this summary, and prepare a similar review after each subsequent chapter.

I. Types of Isomers

Recall that **isomers** are compounds which have the same molecular formula, but different structural formulas (i.e., different molecular structures). As a result of the structural differences, the chemical and physical properties of the isomers are different. A simple example of this relates to compounds of molecular formula $C_2H_6O$. This formula may represent either (a) dimethyl ether or (b) ethanol.

(a) $H_3COCH_3$ (b) $H_3CCH_2OH$

These two isomers have different **structures**, different **types of bonds**, different **functional groups** and correspondingly different chemical and physical properties. These isomers are often referred to as **structural isomers** or functional isomers.

Although structural isomers are in many cases very obvious, other isomer types are more subtle. **Stereoisomers** are compounds which have the same connections of atoms (i.e., each isomer has the same number and types of bonds), but which have different three-dimensional arrangements in space. There are two major classes of stereoisomers of interest to us.

The first type is the **geometric isomer** which results from the restriction of rotation about a bond. The most obvious type of isomers in this case may be the cis- and trans-isomers of a substituted olefin such as $C_2Cl_2H_2$.

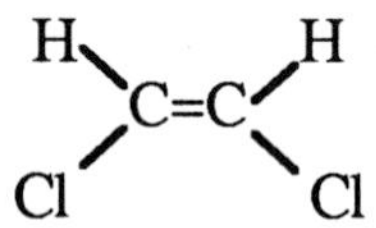

cis-isomer

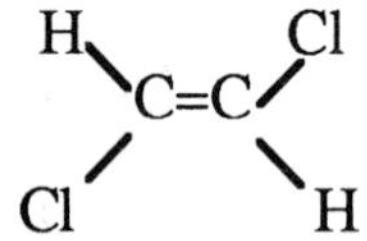

trans-isomer

The second type is the **optical isomer**. Plane polarized light is rotated on passing through a solution of an optical isomer. Two subdivisions of optical isomers are (a) enantiomers and (b) diasteriomers.

In summary, the isomeric forms of molecules can be subdivided into groups as shown below. Much of our interest will focus on optical isomers since monosaccharides, amino acids, nucleotides, lipids and the major macromolecules such as polysaccharides, proteins and nucleic acids are all optically active.

## Isomer Classification

| Isomer Type | Character | Example |
|---|---|---|
| Structural | Same molecular formula, but different molecular structure | $H_3CCH_2OH$<br>$H_3COCH_3$ |
| Stereoisomers | Same connectivity of atoms, but different three-dimensional arrangement in space | |
| a. Geometric | cis- or trans-arrangement about a carbon-carbon double bond, such as observed in some fatty acids (Chapter 17) | cis- and trans-isomer of $C_2Cl_2H_2$ |
| b. Optical | | |
| (i) Enantiomers | Nonsuperimposable mirror images | D- and L-forms of glyceraldehyde (refer to Fig. 16.11 in text) |
| (ii) Diastereomers | Optical isomers which are not enantiomers | α- and β-forms of D-glucose |

## II. Structures and Reactions

The cyclic forms of monosaccharides contain a **hemiacetal** unit formed by the internal reaction of the alcohol group reacting with the carbon atom of the adehyde group. Disaccharides contain an **acetal** unit and, if it includes a reducing sugar unit, it will also contain a hemiacetal unit. The formation and stability of these saccharide units is a major focus of this chapter. As introduced in Chapter 13, the reaction of an aldehyde or a ketone with an alcohol yields the formation of a hemiacetal or a hemiketal, respectively.

a. 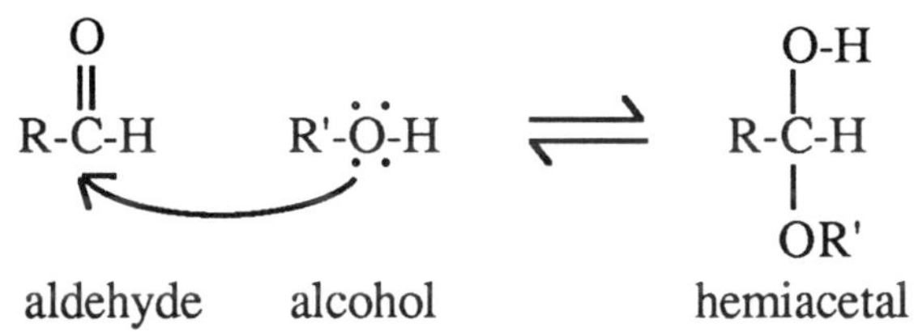


b. 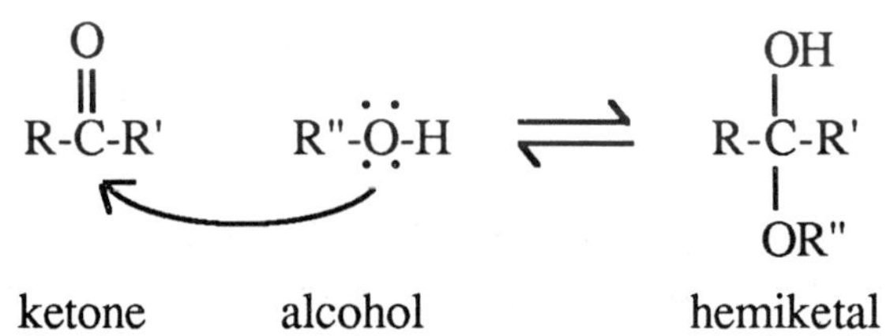


Recall that hemiacetals or hemiketals are unstable unless cyclic systems are formed as in the 5- or 6- membered furanose or pyranose rings, respectively. For D-glucose, note the following:

(i) the highest numbered <u>chiral</u> carbon on the open-chain defines whether the monosaccharide is in the D- or L-series.

(ii) the functional groups in the molecule

(iii) the equilibrium between the open-chain and the cyclic forms

(iv) conversion of the open chain to the cyclic form produces an additional chiral carbon

(v) the position (i.e., the atom number) of the reacting groups in the open-chain form and the reactive group in the products

(vi) the orientation of the groups on C-1 in the cyclic form

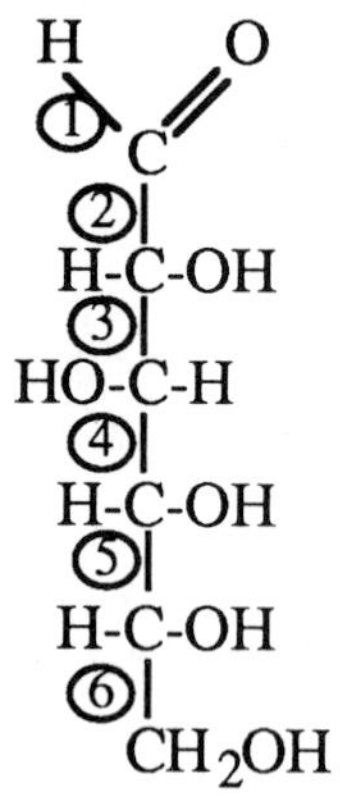

(open-chain form)

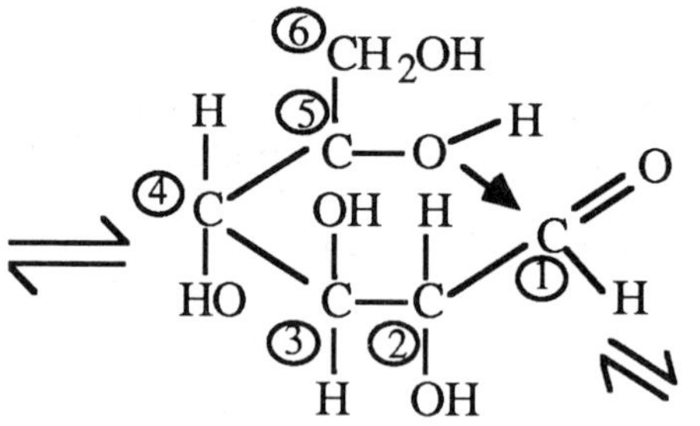

(open-chain form drawn in conformation which leads to internal hemiacetal formation)

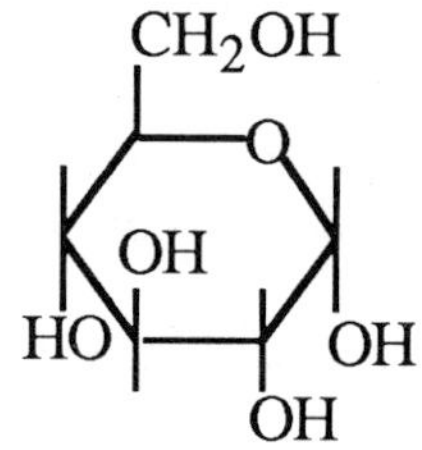

α-D-glucose

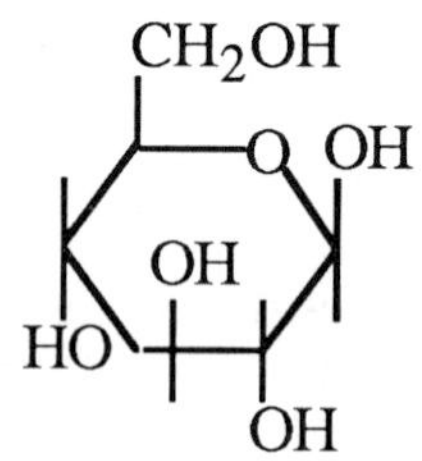

β-D-glucose

Note that the anomeric carbon is at the C-1 position in the pyranose ring. As an exercise, draw the structures for D-fructose in an analogous reaction involving the formation of a five-membered ring containing a hemiketal unit.

Two reactions to review are: (1) the oxidation of mono- or disaccharides which contain a hemiacetal or hemiketal unit and (ii) the hydrolysis reaction of the acetal linkages in polysaccharides which produces monosaccharides. The hydrolysis reaction is catalyzed by acids and enzymes in the body.

## SELF-TEST QUESTIONS

MULTIPLE CHOICE. Select the correct answer from the choices listed. In some cases, more than one answer is correct.

1. Which of the following properties are associated with D-fructose?

   a. optically active
   b. undergoes mutarotation
   c. disaccharide
   d. rotates plane-polarized light

e. is or contains a reducing sugar
f. contains an $\alpha$ (1--->4) linkage

2. Enantiomers have which of the following characteristics?

   a. rotate ordinary light
   b. have the same melting point
   c. react with optically active molecules at the same rate
   d. are superimposable mirror images

3. Starch has which of the following characteristics?

   a. $\beta$ (1--->4) glycosidic linkages
   b. made up of amylose and glycogen
   c. is a monosaccharide
   d. optically active

4. Which of the following molecules are disaccharides?

   a. heparin
   b. sucrose
   c. maltose
   d. acetone

5. The specific rotation of a molecule will depend on which of the following properties?

   a. molecular weight of the molecule
   b. path length of the light
   c. temperature
   d. wavelength of polarized light used

<u>COMPLETION.</u> Complete the following statements with the appropriate word, phrase or number.

1. The process in which the $\alpha$- and $\beta$-ring forms of a monosaccharide interconvert by way of the open chain structure is called ______________.

2. Photosynthesis is the process in which plants combine ______________ and ______________ to produce ______________. Both ______________ and ______________ are required for this reaction.

3. The hydrolysis of sucrose produces ______________ and ______________.

4. The polysaccharide which contains the $\beta$ (1--->4) glycosidic linkage is ______________.

5. The specific rotation for the optically active molecule X is + 6.0°. The solution contained 2.5 g X per cc and the measurement was carried out in a sample tube 20.0 cm long. The rotation observed in the measurement was ______________.

6. A pair of diastereomers are ______________ which are not ______________. The melting points for these two isomers are ______________.

7. Amylopectin differs from amylose in that it contains ______________ which involves ______________ glycosidic linkages.

8. Blood type ________ is the universal donor because it has ______________ of the blood ______________ A or B.

9. The number of chiral carbons in aldoheptose is ______________; therefore, the maximum number of stereoisomers is ______________.

10. Lactose is an example of a ______________ disaccharide because it contains a ______________ unit; therefore it (will or will not) undergo mutarotation.

11. Two common features of heparin and hyaluronic acid are that they both contain ______________ and ______________.

12. The ______________ form of tartaric acid is optically ______________ and is a ______________ of the enantiomeric pair ((+) and (-) forms) of isomers.

## ANSWERS TO SELF-TEST QUESTIONS

Multiple Choice

1. a, b, d, e
2. b
3. d
4. b, c
5. b, c, d

Completion

1. mutarotation
2. $H_2O$, $CO_2$, carbohydrates; light ; chlorophyll
3. glucose, fructose
4. cellulose
5. +30°
6. optical isomers, enantiomers, different
7. chain branching, (1-6)
8. 0, neither, antigens
9. 5, $2^5 = 32$
10. reducing, hemiacetal, will
11. pyranose rings, acidic side groups
12. meso, inactive, diastereomer

It takes a membrane to make sense
out of disorder in biology.
- Lewis Thomas
The Lives of a Cell

# 17

# Lipids

## CHAPTER OBJECTIVES

Lipids are defined **operationally** - that is, those molecules in a cell which can be extracted with non-polar solvents are collectively classified as lipids. This classification procedure is not very rigorous and as a result a rather diverse group of molecules are enveloped within the definition. The focus of this chapter is to define the (1) COMPOSITION, (2) STRUCTURE, (3) PROPERTIES and (4) FUNCTION of these molecules in the cell. Keep this in mind as you review the material. As you will see, the functions range from providing the basic structure of all membranes, to highly specific hormones, some of which regulate metabolism, while others are very specific and define our sex.

After you have studied the chapter and worked the assigned exercises in the text and the study guide, you should be able to do the following.

1. Name the four main subclasses of lipids and give a few examples of each.
2. Draw the general structure of a mono-, di- and triglyceride (all fats) which includes saturated and/or unsaturated fatty acids.
3. Explain how the unsaturation in fatty acids influences their melting point.
4. Outline the Iodine Number test, which is used to determine the extent of unsaturation in a fat or oil.
5. Write out the general equations for the (1) hydrogenation of unsaturated fats or oils and for the (2) saponification of all fats or oils.
6. Define the lipid bilayer and describe the fluid mosaic model of a membrane.
7. Draw a phosphoglyceride and a sphingolipid and indicate their main function.

8. Draw the basic steroid ring system and the structure for cholesterol. List the molecules produced upon oxidation of cholesterol.

9. Outline the interrelationship between HDLs, LDLs, cholesterol and its metabolism and their possible roles in the development of atherosclerosis.

10. Outline the details of how bile salts and lipases work in concert to break down lipids.

11. Draw the structure of arachidonic acid and point out some common structural changes that occur on production of prostaglandins.

12. Define the important terms and comparisons in this chapter and give specific examples where appropriate.

## IMPORTANT TERMS AND COMPARISONS

Hydrophobic and Hydrophilic
Fat, Oils and Waxes
Glycerol
Fatty Acid
Van der Waal Force
Ester Linkage
Cis Isomer
Essential Fatty Acid
Polyunsaturated
Hydrogenation
Rancid
Saponification
Soap and Detergent
Membrane and Lipid Bilayer
Phospholipid and Glycolipid
Phosphoglyceride
Choline
HDLs and LDLs
Phosphatidyl Choline or Lecithin
Cephalin and Lecithin
Sphingosine
Sphingomyelin
Tay-Sachs Disease
Steroid and Cholesterol
Progesterone
Mineralcorticoid
Glucocorticoid
Aldosterone and Cortisol
Bile Salts
Sex Hormones
Testosterone
Estradiol
Oral Contraceptives
Lipase
Prostaglandins and Leukotrienes

## SELF-TEST QUESTIONS

MULTIPLE CHOICE. In the following exercises, select the correct answer from the choices listed. In some cases, two or more choices will be correct.

1. Fats differ from waxes in that fats have:

   a. more unsaturation
   b. a glycerol backbone
   c. higher melting points
   d. longer fatty acids

2. Oleic acid is an unsaturated fatty acid containing:

   a. 12 carbons
   b. 14 carbons
   c. 16 carbons
   d. 18 carbons

3. Both stearic acid and linoleic acid have 18 carbons; however, linoleic acid is unsaturated, while stearic acid is saturated. The melting point of stearic acid:

a. is higher
b. is lower
c. is the same as linoleic acid
d. relative to that for linoleic acid cannot be predicted

4. Liquid soaps are generally salts of fatty acids in which the cation is:

a. $K^+$
b. $Na^+$
c. $Ca^{+2}$
d. $Mg^{+2}$

5. Which of the following structures represents cholesterol?

a.

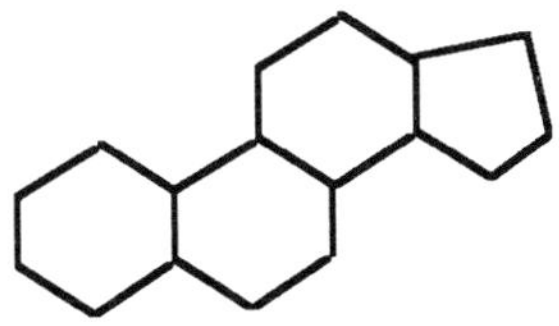

b.

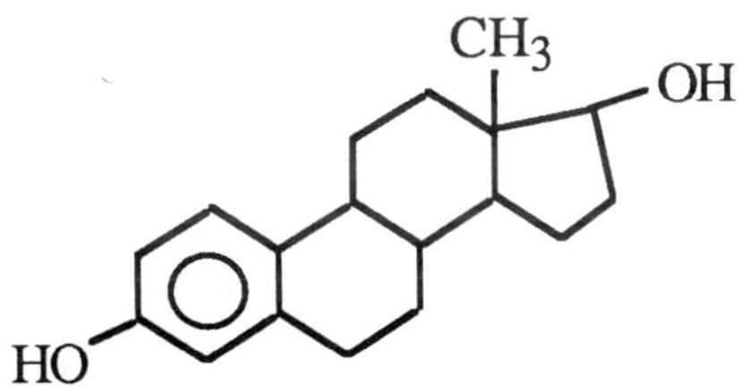

c.

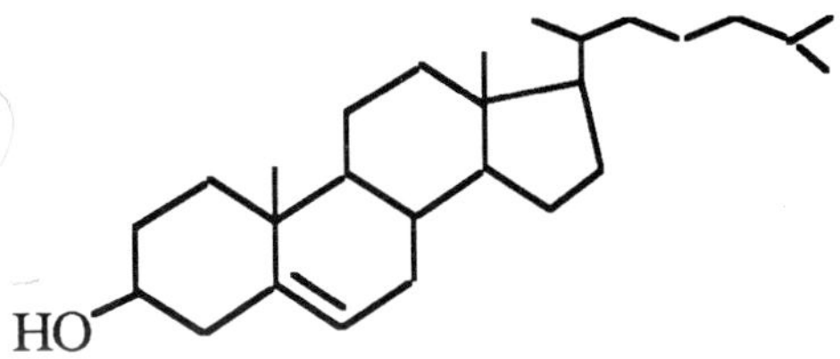

d.

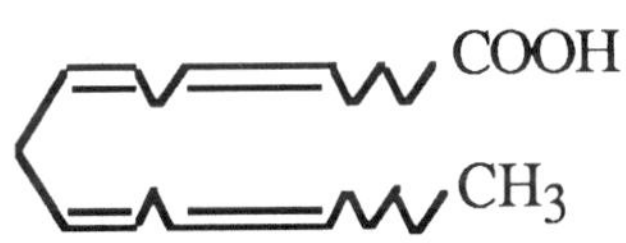

6. Prostaglandin group, $PGE_2$, possesses:

a. ring closure at C-8 and C-12
b. 20 carbon atoms
c. a carbonyl group at C-9
d. two double bonds

7. Multiple sclerosis is thought to be associated with:

a. degradation of the myelin sheath
b. inhibited prostaglandin synthesis
c. lipid storage
d. a deficiency of enzyme 21 hydroxylase

8. Lipid bilayers are composed mainly of:

a. complex lipids
b. steroids
c. prostaglandins
d. fatty acids

9. Basic hydrolysis or saponification of triglycerides will produce:

a. glycerol
b. fatty acids
c. salts of fatty acids
d. bile salts

10. The 18 carbon backbone of sphingosine is found in:

a. sphingomyelin
b. myelin sheath
c. phosphatidyl choline
d. testosterone

11. Oils are:

a. liquid fats
b. completely saturated
c. high molecular weight fatty acids
d. colored

12. Derivatives of what molecules are effective oral contraceptives?

a. progesterone
b. estradiol
c. $PGF_{2a}$
d. cortisone

13. Low density lipoprotein (LDL) has which of the following characteristics:

a. Contains more cholesterol than high density lipoprotein (HDL).
b. Soluble in the blood plasma
c. Carries cholesterol to the cells
d. Binds to receptor molecules on the cell surface within the coated pits.

<u>COMPLETION</u>. Write the word, phrase or number in the blank space in answering the question.

1. Energy is stored in our bodies in the form of ______________ because burning it produces twice as much energy as the burning of ______________.

2. Aspirin has been found to inhibit ______________ synthesis.

3. ______________, which has 20 carbon atoms, is the starting material in the body for prostaglandin synthesis.

4. The enzymes involved in lipid metabolism are called ______________.

5. A widely known disease associated with lipid storage is ______________.

6. ______________ are negatively charged salts derived from the oxidation of ______________.

7. The hydrophobic segment of a detergent is associated with the ______________.

8. All unsaturated fatty acids exist as ______________ at room temperature.

9. Rancidity is caused in part by the ______________ of the fat or oil to aldehydes or by the production of ______________ by the hydrolysis of triglycerides.

10. The fluid mosaic model of membranes proposed that _______________ molecules are floating on and within the _______________.

11. For a series of fatty acids with the same number of carbon atoms, those with the highest iodine number have the greatest degree of _______________.

12. The common name for phosphatidyl cholines is _______________.

13. Leukotrienes are _______________ of hormonal responses. Although similar in structure to prostaglandins, there is no __________ __________ in the structure of leukotrienes.

14. A faulty cholesterol transport system is often characterized by high levels of __________ and low levels of __________, many times resulting from not enough functional __________ __________.

## ANSWERS TO SELF-TEST QUESTIONS

Multiple Choice

1. b
2. d
3. a
4. a
5. c
6. a, b, c, d
7. a
8. a
9. a, c
10. a
11. a
12. a
13. a, b, c, d

Completion

1. fat, carbohydrates
2. prostaglandin
3. Arachadonic acid
4. lipases
5. any of the 5 listed in Box 17.F in the text
6. Bile salts, cholesterol
7. non-polar hydrocarbon chain
8. liquids
9. oxidation, fatty acids
10. protein, lipid bilayer
11. unsaturation
12. lecithin
13. mediators, ring closure
14. LDL, HDL, LDL receptors

"Isn't it astonishing that all these secrets have been preserved for so many years just so that we could discover them."
- Orville Wright
Miracle at Kitty Hawk

# 18

# Proteins

## CHAPTER OBJECTIVES

This chapter, similar to the previous chapter on lipids, concentrates on the (1) COMPOSITION, (2) SEQUENCE, (3) the levels of STRUCTURE and the (4) diverse FUNCTIONS of proteins. Proteins and/or enzymes are associated with, in some way, virtually all reactions in the cell and therefore will be of interest in all aspects of cellular regulation and metabolic processes.

After you have studied the chapter and worked the assigned exercises in the text and the study guide, you should be able to do the following.

1. List six major functions of proteins and give an example of a protein which is involved in each biological function.
2. Write out the structures for at least two amino acids which are representative of an amino acid with a (1) nonpolar, (2) polar, but neutral, (3) acidic and (4) basic R group. Also indicate the unique character of the amino acids glycine, proline and cysteine.
3. Demonstrate the amphoteric character of amino acids by writing the reactions for the zwitterion with an (1) acid or (2) base.
4. Write out the reversible reaction for the oxidation of cysteine and indicate how the production of cystine in a **protein** might stabilize the tertiary structure of proteins.
5. Draw the structure for a tripeptide, pointing out (1) the peptide linkages, (2) the chiral carbon atoms in the backbone, (3) the N- and C-terminal ends and (4) the R groups.
6. Name the four levels of structure in proteins and briefly describe what each is associated with.
7. Indicate the four forces which are involved in determining protein structure and draw a representation of each.

8. Characterize the secondary structures in α-helix and β-pleated sheet in terms of structure and the types of forces involved in each.

9. Indicate what changes occur in proteins by the addition of specific denaturation agents and list some practical applications of denaturation.

10. Define the important terms and comparisons in this chapter and give specific examples where appropriate.

## IMPORTANT TERMS AND COMPARISONS

Catalyst
Enzyme
Fibrous and Globular Protein
R Group or Side Chain
Essential Amino Acid
Complete Protein (in the dietary sense)
Kwashiorkor
Enantiomers
Racemization
D- and L- Structural Forms
Amphoteric
Buffer Solution
Cysteine and Cystine
Glutathione
Di-, Tri- and Polypeptide
Protein and Peptide
Amino Acid Residue
N-Terminal and C-Terminal Ends
Peptide Backbone
Primary, Secondary and Tertiary Structure
Sickle Cell Anemia and Trait
Homo- and Heterozgote
Intra- and Intermolecular Hydrogen Bonding
α-Helix, β-Pleated Sheet and Random Coil
Tropocollagen and Collagen
Disulfide Bridge
Hemoglobin
Globin
Gelatin

## SELF-TEST QUESTIONS

MULTIPLE CHOICE. In the following exercises, select the correct answer from the choices listed. In some cases, two or more choices will be correct.

1. The number of tetrapeptides which can be made from any of the 20 amino acids is:

   a. $1.6 \times 10^5$
   b. 20
   c. 4
   d. 80

2. Glycine is an unique amino acid because it:

   a. has no chiral carbon
   b. has a sulfur containing R group
   c. cannot form peptide bond
   d. is an essential amino acid

3. The α-helix is a common form of:

   a. primary structure
   b. secondary structure
   c. tertiary structure
   d. a denatured region

4. Which of the following is a transport protein?

a. hemoglobin
b. pepsin
c. collagen
d. oxytocin

5. A zwitterion has which of the following characteristics?

a. a high melting point
b. no net charge
c. soluble in water
d. all of above

6. Arginine is a basic amino acid while alanine is nonpolar. The isoelectric point for arginine is:

a. higher than for alanine
b. lower than for alanine
c. about the same as alanine
d. cannot predict

7. Insulin is a hormone made up of two chains crosslinked by two disulfide bonds. This protein has:

a. 2 subunits
b. 1 subunit
c. no subunits
d. 4 subunits

8. Which of the following R groups associated with an amino acid can take part in hydrogen bonding?

a. $-CH_2SCH_3$(methionine)
b. $-CH_2OH$ (serine)
c. $-CH_2CO_2H$ (aspartic acid)
d. $-CH_2CH_2CH_2CH_2NH_3^+$ (lysine)

9. Hydrophobic interactions occur between the R groups with:

a. acidic character
b. nonpolar character
c. basic character
d. all of above

10. The number of amino acid residues in the cyclic nonapeptide vasopressin is:

a. 7
b. 9
c. 19
d. 90

<u>COMPLETION</u>. Write the word, phrase or number in the blank space in answering the question.

1. In the tripeptide drawn below, circle and clearly label the following units: peptide linkages; C- and N-terminal ends; chiral carbons; R groups.

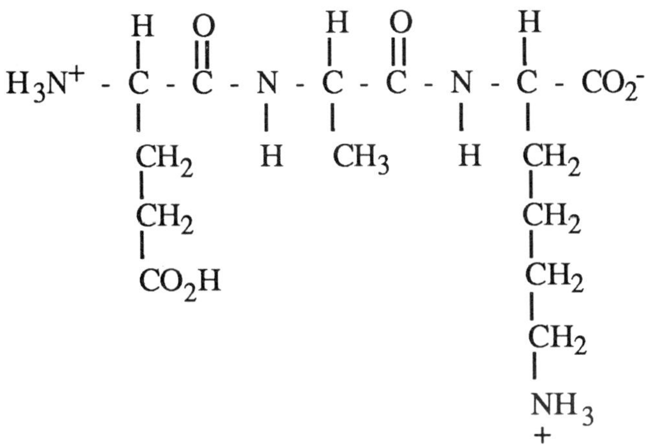

2. Tropocollagen is the _____________ form of collagen and is made up of _____________ _______________ units. Insoluble collagen results from covalent _______________ of two _____________ residues on adjacent chains in the helix.

3. A covalent bond in the _______________ _______________, is the force which is associated with the _______________ structure of a protein.

4. Agents which can reversibly denature a single subunit protein do this by breaking forces involved in the _______________ and _______________ structure.

5. Sickle cell trait occurs in people who have _______________ gene(s) programmed to produce sickle cell hemoglobin.

6. The α-helix is stabilized by _____________-molecular hydrogen bonding, while the β-pleated sheet structure is stabilized by _____________-molecular hydrogen bonding. The interaction is in both cases between the _______________ and the _______________ groups.

7. Of the twenty amino acids, only _____________ of them are termed essential amino acids because they cannot be synthesized in sufficient quantities within the body.

8. Complete the following reactions:

a. 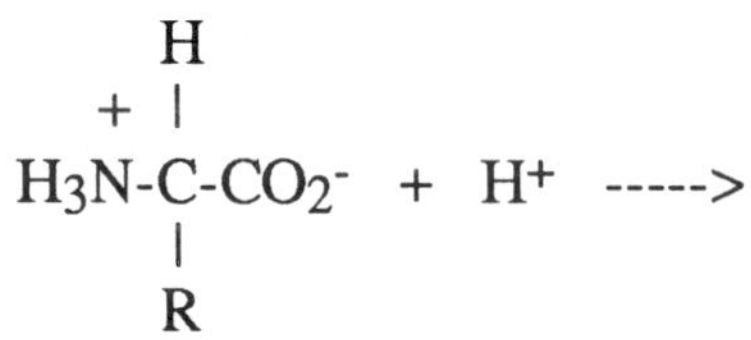 _______________________________

b. 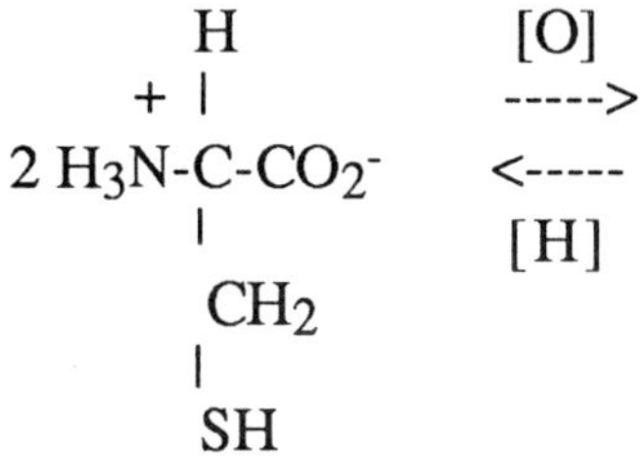 _______________________________

c. tetrapeptide + HCl $\xrightarrow{\text{heat}}$ _______________________________

d. 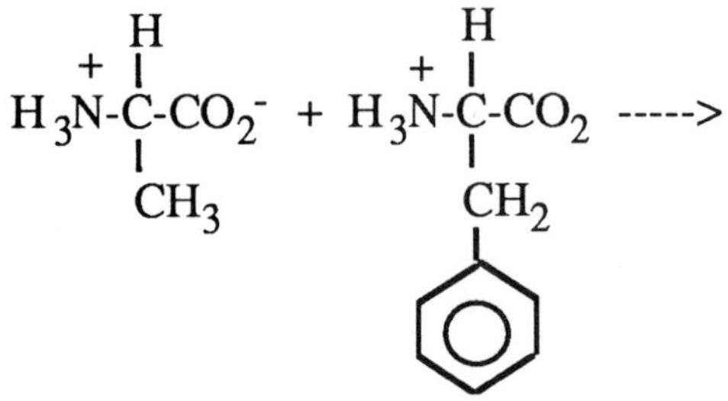
$H_3\overset{+}{N}$-CH($CH_3$)-$CO_2^-$ + $H_3\overset{+}{N}$-CH($CH_2C_6H_5$)-$CO_2$ -----> ______________________ + ______________________

alanine phenylalanine

9. ______________ structure occurs only in proteins which have ______________ subunits, such as in hemoglobin.

10. Kwashiorkor is a ______________ ______________ disease which is widespread in many underdeveloped countries.

## ANSWERS TO SELF-TEST QUESTIONS

Multiple Choice

1. a
2. a
3. b
4. a
5. d
6. a
7. b (Figure 18.3, text)
8. b, c, d
9. b
10. b

Completion

1.

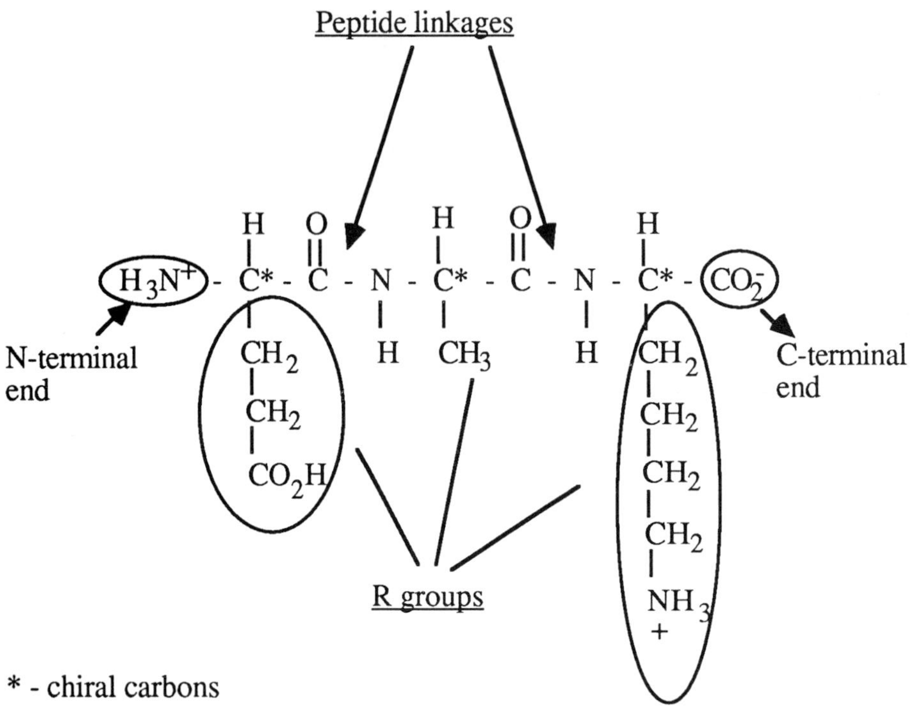


* - chiral carbons

2. soluble; triple helical; crosslinking; lysine
3. peptide linkage; primary
4. secondary; tertiary
5. one
6. intra; inter; 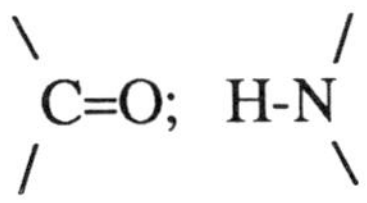
7. 10
8.

a. 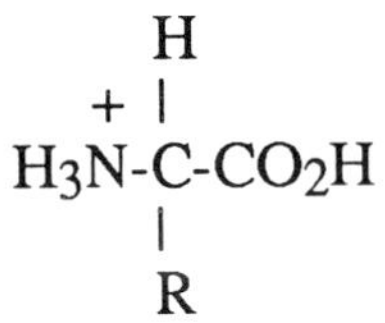

b. 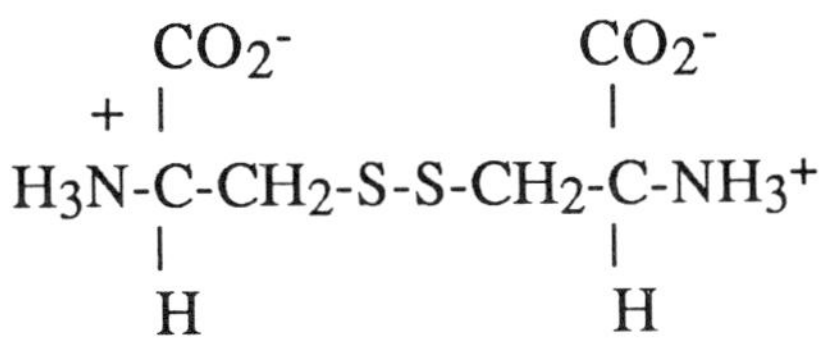

cystine, containing a disulfide bond

c. 4 amino acids

d. 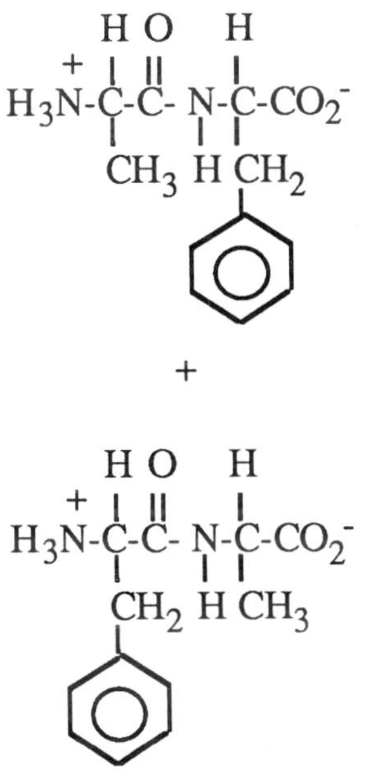

9. Quaternary, multiple (2 or more)
10. protein deficiency

Were it not equipped with catalysts, every living unit would be a static system

F.G. Hopkins
Science, 78, 219 (1933)

# 19

# Enzymes

## CHAPTER OBJECTIVES

After you have studied the chapter and worked the assigned exercises in the text and the study guide, you should be able to do the following.

1. Explain the statement that enzymes increase the rate of the reaction, but do not change the position of the equilibrium.

2. List the six major categories of enzymes and generally indicate the type of reaction that is involved.

3. Distinguish between the terms enzyme, cofactor, coenzyme, apoenzyme, proenzyme and active site.

4. Describe, in both words and in figures, the effect the following have on the rate of a catalyzed reaction:

   enzyme concentration
   substrate concentration
   temperature
   pH
   inhibitors

5. Outline and compare the proposals in the (1) lock and key model and the (2) induced fit model to explain enzymatic action and specificity.

6. Describe how a competitive and a non-competitive inhibitor produce their effect.

7. Define active site and list factors associated with it which are responsible for its specificity toward molecular substrates.

8. Explain how the activity of an allosteric enzyme can be modulated in a positive or negative way by regulator interactions. Relate this effect to the feedback control mechanisms found in complex reaction systems.

9. Describe how penicillin is used for bacterial infections and explain the role of clavulinic acid in killing penicillin resistant bacteria.

10. Define the important terms and comparisons in this chapter and give specific examples where appropriate.

## IMPORTANT TERMS AND COMPARISONS

Enzyme and Catalyst
-ase
Oxidoreductase
Transferase
Hydrolase
Lyase
Isomerase
Ligase
Cofactor and Coenzyme
Substrate and Inhibitor
Proenzyme
Reversible Inhibitor
Competitive and Non-Competitive Inhibitor
Irreversible Inhibitor
Sulfa Drugs
Negative and Positive Modulation
Penicillin

## SELF-TEST QUESTIONS

MULTIPLE CHOICE. In the following exercises, select the correct answer from the choices listed. In some cases, two or more choices will be correct.

1. A molecule which is structurally similar to the substrate for an enzyme will probably be a:

   a. competitive inhibitor
   b. cofactor
   c. regulator
   d. noncompetitive inhibitor

2. The site on an allosteric enzyme which is directly involved in modulation is called:

   a. the prosthetic group
   b. regulatory site
   c. active site
   d. target site

3. The regulatory sequence of reactions shown below uses a feedback mechanism of inhibition. The last product, D, in the sequence usually:

   a. inhibits $E_1$
   b. inhibits $E_2$
   c. inhibits $E_3$
   d. inhibits all enzymes

$$A \xrightarrow{E_1} B \xrightarrow{E_2} C \xrightarrow{E_3} D$$

4. Competitive inhibition can be overcome by:

   a. increasing substrate concentration
   b. increasing pH
   c. decreasing temperature
   d. all of above

5. The pH influences most enzymatic reactions because pH changes can:

a. hydrolyze the protein
b. produce protonation or deprotonation of essential amino acid residues in the active site
c. change its primary structure
d. effect the optical activity of the protein

6. A significant increase in the temperature of an enzymatic reaction will reduce the rate because the:

a. enzyme acts on itself as an inhibitor
b. protein is partially or completely denatured
c. protein undergoes hydrolysis

7. Consider the possible shapes that a molecule might have to act as a competitive inhibitor to the reaction shown below:

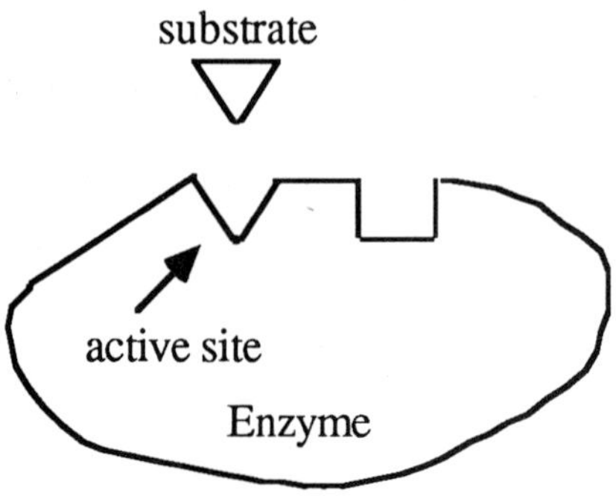


Indicate which of the following molecules (molecular shapes) could act as a competitive inhibitor.

a. 

c. 

b. 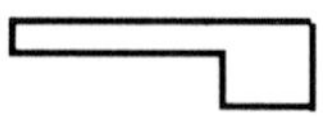

d. 

8. The apoprotein corresponds to which part of an enzyme?

a. coenzyme
b. protein portion
c. cofactor
d. part cleaved off in a proenzyme

9. The site on the enzyme at which the substrate interacts is called the

a. regulatory site
b. modulator site
c. active site
d. allosteric site

                                      ----->
10. Consider the reaction A + B <----- C + D. An enzyme used in this reaction will:

   a. increase the equilibrium constant
   b. increase the amount of C and D produced
   c. increase the rate at which C and D are produced
   d. lower the activation energy for both the forward and reverse reaction

11. An allosteric enzyme has which of the following characteristics?

   a. A modulator can also bind to the active site
   b. The binding of a modulator molecule to the regulatory site always inhibits enzymatic activity
   c. The binding of the modulator to the regulatory site can either increase or decrease the enzymatic activity
   d. The active site and the regulatory sites are always in different locations and cannot be identical.
   e. In most cases, the allosteric enzyme is a multisubunit protein

<u>Completion</u>. Write the word, phrase or number in the blank space in answering the question.

1. Classify each of the following enzymes into one of the six major classes.

| | | <u>Class</u> |
|---|---|---|
| a. | Ribonuclease | |
| b. | Alcohol Dehydrogenase | |
| c. | DNA Topoisomerase | |
| d. | Pepsin | |
| e. | Alanine Racemase | |
| f. | Pyruvate Kinase | |
| g. | Acetylcholinesterase | |
| h. | Penicillinase | |

2. Lipases are enzymes which are (<u>more or less</u>) specific than trypsin.

3. Of the two main models used to explain enzymatic action, the ________________ model is regarded as more flexible than the ________________ model.

4. A number of enzymes require a non-protein component called a ________________ or ________________ to exhibit enzymatic activity.

5. Trypsinogen is an example of a ________________ which can only be ________________ by removal of a portion of the protein.

6. If the enzyme concentration as an enzymatic reaction is tripled, the reaction rate will ________________.

7. Heavy metals, such as $Hg^{+2}$ or $Pb^{+2}$, act as ________________ inhibitors to many enzymes.

8. Label the following plot to show a pH profile for an enzyme in which the highest activity occurs at pH 4.5 and decreases to nearly zero at pH 7.0 and 2.0.

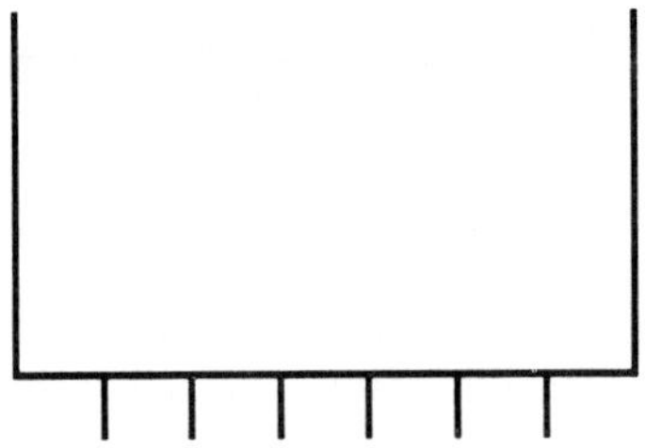

9. Draw the substrate, enzyme and enzyme-substrate (E-S) complex as they may be viewed in the induced fit model.

-------->
<--------

_____________ + _____________ <--------> ____________________

Substrate + Enzyme (E-S) complex

10. ______________ ______________ is administered with penicillin to act as an irreversible inhibitor of ______________ in penicillin resistant cells.

## ANSWERS TO SELF-TEST QUESTIONS

Multiple Choice

1. a
2. b
3. a
4. a
5. b
6. b
7. a, b (note that the left side of b will inhibit substrate binding)
8. b
9. c
10. c, d
11. c, d, e

Completion

1. a. Hydrolase
   b. Oxidoreductase
   c. Isomerase
   d. Hydrolase
   e. Isomerase
   f. Transferase (specific for phosphates)
   g. Hydrolase
   h. Hydrolase
2. less
3. induced fit; lock and key
4. cofactor, coenzyme
5. proenzyme, activated
6. triple
7. irreversible

8.

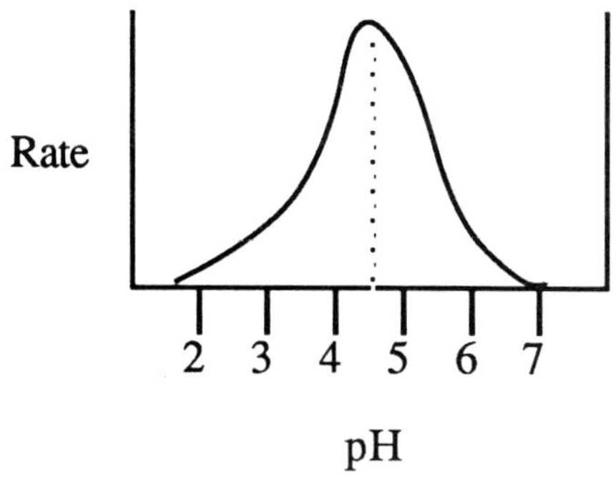


9.

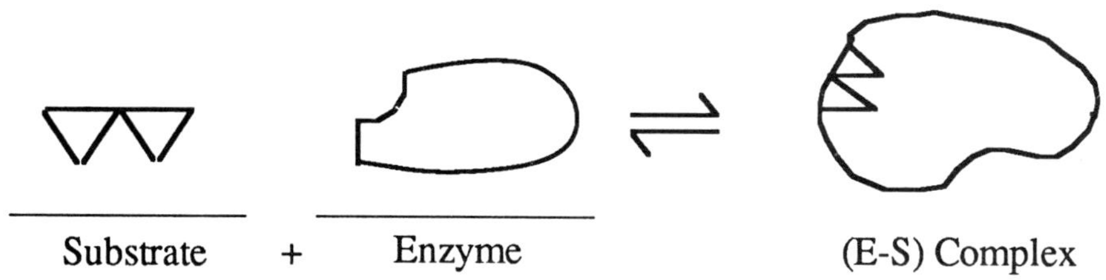


10. Clavulinic acid, penicillinase

`Whatever a cell does has to be paid
for in the currency of energy
- Szent-Gyorgyi

# 20

# Bioenergetics: How the Body Converts Food to Energy

## CHAPTER OBJECTIVES

After you have studied the chapter and worked the assigned exercises in the text and the study guide, you should be able to do the following.

1. List the (molecular) components contained in ATP, the coenzymes $NAD^+$ and FAD, and acetyl CoA. Write out a general reaction in which each takes part and draw the part of each structure which is changed as a result of the reaction.

2. Distinguish generally between the catabolic and the anabolic process in terms of the consumption or production of ATP, $NAD^+$ and FAD.

3. Draw a clear schematic representation of a mitochondrion pointing out the outer membrane, inner membrane, intermembrane region, the matrix, cristae, and the location of the proton translocating ATP-ase.

4. State the major role of ATP in metabolic reactions.

5. Explain the role of $NAD^+$ and FAD in the citric acid cycle and in the electron transport chain.

6. Describe the role of acetyl CoA in the Krebs cycle.

7. Name the major species in the Krebs cycle.

8. Indicate the type of reaction involved at each step in the Krebs cycle and indicate (in words) an example of each.

9. Explain how the reduced coenzymes, NADH and $FADH_2$, are involved in oxidative phosphorylation and why each produces a different number of ATP molecules.

10. Determine the number of ATP molecules produced in the Krebs cycle and oxidative phosphorylation as a result of the oxidation of one acetyl group.

11. Describe the chemiosmotic hypothesis.

12. Describe the function of the proton translocating ATP-ase.

13. Explain the value of using an uncoupling agent in oxidative phosphorylation studies.

14. Give specific examples of the use of ATP to produce mechanical, electrical and chemical energy.

15. Define the important terms and comparisons in this chapter and give specific examples where appropriate.

## IMPORTANT TERMS AND COMPARISONS

Catabolism and Anabolism
Activated $C_2$ Fragments or Acetyl Coenzyme A
Organelle
Mitochondrion-Outer and Inner Membrane, Intermembrane Space, Matrix and Cristae
Krebs Cycle, Tricarboxylic Acid Cycle and Citric Acid Cycle
AMP, ADP, ATP
Adenosine and Adenine
Ribose
Phosphoric Anhydride Linkage
Phosphate Ester Linkage
High Energy Phosphate Bond
Inorganic Phosphate
Turnover Rate
Coenzyme
$NAD^+$ and NADH
FAD and $FADH_2$
Nicotinamide
Riboflavin and Flavin
Nicotinic Acid
Niacin and Niacinamide
Pantothenic Acid
Mercaptoethylamine
Oxaloacetate
Dehydration Reaction
Tertiary and Secondary Alcohol
Oxidative Decarboxylation
Guanine
GTP
Flavoprotein
Q Enzyme
Cytochromes
Proton Translocating ATP-ase
Uncoupling Agent
Protonphore
Myosin and Actin

## I. Stages of Metabolism

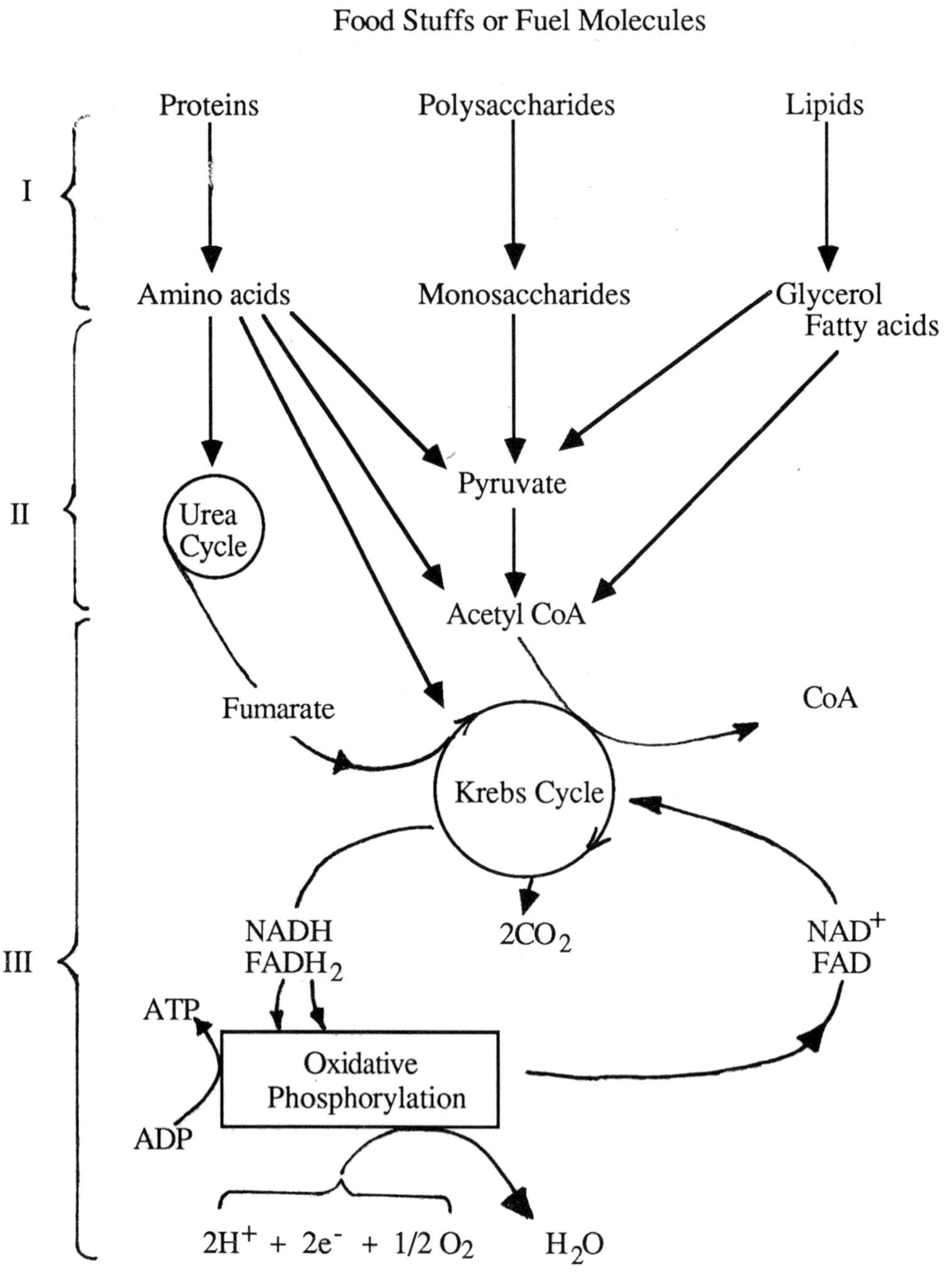


In the initial stage, the macromolecules in food are digested to monomers. Proteins are degraded to amino acids, polysaccharides to glucose and other monosaccharides and lipids are hydrolyzed to fatty acids and glycerol. In the second stage, the sugars, fatty acids, glycerol and many of the amino acids are converted into the activated acetyl units ($C_2$ fragments) of acetyl coenzyme A. The nitrogens in amino acids are processed in the urea cycle, while the deaminated carbon skeletons are processed directly or in a few steps into the Krebs cycle. The major focus in stage III involves the Krebs cycle and oxidative phosphorylation. In these final stages of the oxidation of "fuel" molecules, the $C_2$ fragment of acetyl CoA enters the Krebs cycle and is completely oxidized to $2CO_2$, reduced coenzymes NADH and $FADH_2$ are produced, along with one GTP molecule.

The reduced coenzymes NADH and $FADH_2$ enter in the electron transport chain and are oxidized to $NAD^+$ and FAD, which are then available for recycling in the Krebs cycle. As a result of the oxidation, hydrogen ions are expelled from the mitochondria and upon re-entry produce ATP from the phosphorylation of ADP. The electrons and the hydrogen ions now combine with $O_2$ to form water. Note that most of the ATP generated from the catabolic metabolism of the foodstuffs is formed in stage III.

## II. Summary of Reactions

a. Interconversion of ATP to ADP + Pi

$$ATP \rightleftharpoons ADP + Pi$$

b. Half reactions for coenzymes $NAD^+$ and FAD

$$NAD^+ + H^+ + 2e^- = NADH$$
$$FAD + 2H^+ + 2e^- = FADH_2$$

c. Overall reaction in Krebs cycle

$$H_3CCOOH + 2H_2O + 3NAD^+ + FAD + GDP + Pi \longrightarrow$$

$$2CO_2 + \underbrace{3NADH + FADH_2}_{} + GTP + 3H^+$$

| waste product | to electron transport chain | energy |
|---|---|---|

d. Overall reaction of oxidative phosphorylation

$$NADH + 3ADP + 1/2O_2 + 3Pi + H^+ = NAD^+ + 3ATP + 4H_2O$$
$$FADH_2 + 2ADP + 1/2O_2 + 2Pi = FAD + 2ATP + 3H_2O$$

| from Krebs cycle | recyle back to Krebs cycle | energy |
|---|---|---|

e. Oxidation of a $C_2$ (acetyl) fragment

$$C_2 + 2O_2 + 12ADP + 12\,Pi = 12\,ATP + 2CO_2$$

III. Citric Acid Cycle*

| Step | Reaction | Type |
|---|---|---|
| 1. | Acetyl CoA + oxaloacetate + $H_2O$ -----> citrate + CoA<br>oxaloacetate $(\underline{C}_4\underline{H}_2\underline{O}_5)$; citrate $(\underline{C}_6\underline{H}_5\underline{O}_7)$ | Condensation reaction |
| 2. | Citrate -----> cis-aconitate + $H_2O$ -----> Isocitrate<br>Citrate $(C_6H_5O_7)$ tertiary alcohol; Isocitrate $(C_6H_5O_7)$ secondary alcohol | Isomerization, by way of a dehydration, followed by a hydration |
| 3. | Isocitrate + $NAD^+$ -----> α-ketoglutarate + NADH + $CO_2$<br>Isocitrate $(\underline{C}_6\underline{H}_5\underline{O}_7)$ alcohol; α-ketoglutarate $(\underline{C}_5\underline{H}_4\underline{O}_5)$ ketone | Oxidative decarboxylation |
| 4/5 | α-ketoglutarate + $NAD^+$ + $H_2O$ + GDP + Pi -----> succinate + $CO_2$ + NADH + $H^+$ + GTP<br>α-ketoglutarate $(\underline{C}_5H_4\underline{O}_5)$ ketone; succinate $(\underline{C}_4H_4\underline{O}_4)$ acid | Oxidative decarboxylation |
| 6. | Succinate + FAD -----> Fumarate + $FADH_2$<br>Succinate $(C_4\underline{H}_4O_4)$; Fumarate $(C_4\underline{H}_2O_5)$ | Oxidation (of succinate) |
| 7. | Fumarate + $H_2O$ -----> malate<br>Fumarate $(C_4H_2\underline{O}_4)$; malate $(C_4H_4\underline{O}_5)$ | Hydration |
| 8. | Malate + $NAD^+$ -----> oxaloacetate + NADH + $H^+$<br>Malate $(C_4\underline{H}_4O_5)$ alcohol; oxaloacetate $(C_4\underline{H}_2O_5)$ ketone | Oxidation |

*The Krebs cycle can be regarded as a catalytic cycle in which a 2 carbon fragment enters the cycle in the form of acetic acid (an acetyl group plus the OH from $H_2O$), is oxidized and leaves the cycle as $CO_2$.

The major species in each step of the cycle can be easily remembered by using the following saying as a mnemonic device. "**O**f **c**ourse, **I** **k**now **s**ome **f**amous **m**en." The first letter in each substrate - **o**xaloacetate, **c**itrate, **i**socitrate, α-**k**etoglutarate, **s**uccinate, **f**umarate and **m**alate begins with these same letters as in the saying.

Again, the overall reaction in the Krebs cycle is:

$$CH_3COOH + 2H_2O + 3NAD^+ + FAD + GDP + Pi \text{ -----> } 2CO_2 + 3NADH + FADH_2\ 3H^+ + GTP$$

IV. Mitochondrion Components

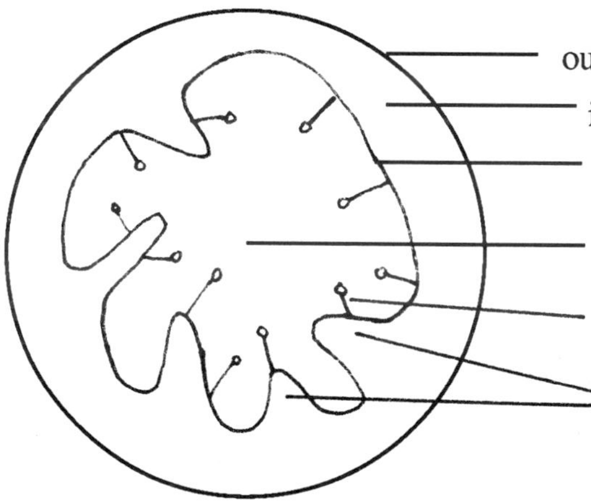


## SELF-TEST QUESTIONS

MULTIPLE CHOICE. In the following exercises, select the correct answer from the choices listed. In some cases, two or more answers will be correct.

1. $FADH_2$ transfers its electrons into the electron transport chain in the mitochondrion directly to

   a. ATP-ase
   b. Flavoprotein
   c. Q Enzyme
   d. Cytochrome c

2. Indicate which of the following processes does not consume ATP.

   a. Muscle contraction
   b. Flow of electrons into mitochondria through proton translocating ATP-ase
   c. Catabolism
   d. Pumping of $K^+$ into the cells by transport proteins

3. In the reaction in which succinate is converted to fumarate in the TCA cycle, succinate undergoes a

a. Hydration
b. Oxidation
c. Isomerization
d. Oxidative decarboxylation

4. Which of the following molecules are produced in the Krebs cycle?

a. $NAD^+$
b. $CO_2$
c. GDP
d. Oxaloacetate

5. The proton translocating ATP-ase is located where in the mitochondrion?

a. Intermembrane space
b. Outer membrane
c. Inside the mitochondrion
d. Outside the inner membrane

6. The number of NADH molecules generated in the Krebs cycle is

a. 1
b. 2
c. 3
d. 4

7. The amount of energy produced by the hydrolysis of ATP is

a. 7.3 kcal/mole
b. 3-4 kcal/mole
c. 32 kcal/mole
d. less than 1 kcal/mole

8. The reduction of FAD to $FADH_2$ occurs in which part of the FAD molecule?

a. Ribose
b. Flavin
c. ADP
d. Nicotinamide

9. Indicate which of the following molecules are carriers of intermediates in metabolism.

a. CoA
b. ADP
c. FAD
d. Nicotinic acid

10. Of the following molecules, which contain high energy bonds?

a. ATP
b. AMP
c. $CO_2$
d. Isocitrate

11. The number of electrons, originating from the electron transport chain, that is needed to produce one mole of water is:

a. 1
b. 2
c. 3
d. 4

12. Niacin is another name for

a. Nicotinic acid
b. α-ketoglutarate
c. Q enzyme
d. Nicotinamide

13. Which of the following terms is NOT another name for the Krebs cycle?

a. Tricarboxylic acid cycle
b. Electron transport chain
c. Oxidative decarboxylation
d. Citric acid cycle

14. The acetyl group in acetyl CoA is linked through what kind of a bond?

a. (C-C) bond
b. (C-S) bond
c. (C-O) bond
d. (C-N) bond

15. Which of the following molecules are NOT an electron carrying enzyme in the mitochondrial membrane?

a. Cytochrome $a_3$
b. Q enzyme
c. Fumarase
d. Citrate synthetase

16. For every two $C_2$ fragments that enter the TCA (tricarboxylic acid) cycle, the number of ATP and (GTP) molecules produced is

a. 24
b. 22
c. 10
d. 12

COMPLETION. Write the word, phrase or number in the blank or draw the appropriate structure in answering the question.

1. A ________________ is a compound that permits $H^+$ ions to pass through the ________________ membrane passively.

2. The correct order of the enzymes, Q enzymes, FeS protein, cytochromes and flavoprotein in the electron transport chain is

_______________ , ________________ , ________________ , ________________

3. Anabolism is associated with the ______________ of molecules while catabolism deals with the ______________ of molecules.

4. ________________ is the molecular unit which is common in FAD, $NAD^+$, ATP and coenzyme A.

5. For each NADH molecule, ________________ pairs of protons are pumped out of the mitochondrion into the ________________ ________________ , with the resultant production of ________________ molecules of ATP.

6. Identify the 6 most important features of the mitochondrion

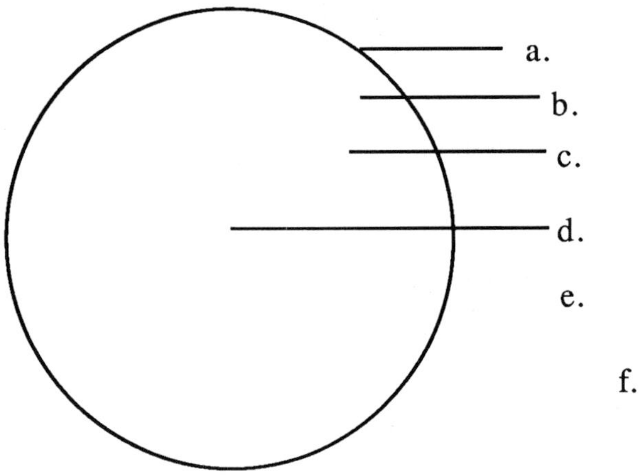


7. Considering the four intermediate carriers, ______________ is a molecular unit unique to FAD, while _______________ is a unique unit occurring in coenzyme A.

8. In the citric acid cycle, the tertiary alcohol in ____________ must be converted to a secondary alcohol in _______________ before the subsequent oxidation to produce a _______________ group can occur.

9. The ________________ hypothesis, proposed by Peter Mitchell states (in a short paragraph).

10. The reactions of the Krebs cycle occur in the ______________ .

11. The products of the reaction of acetyl CoA and oxaloacetate are ______________ and ________________ .

12. The reactions in the electron transport chain eventually lead to the reaction of $O_2$ with _______________ and _______________ .

13. $NAD^+$ is a coenzyme which can act as an _____________ agent in the Kreb's cycle. In the process of being converted to the NADH form, it becomes __________ and picks up __________ proton and ________ electrons.

## ANSWERS TO SELF-TEST QUESTIONS

Multiple Choice

1. c
2. b, c
3. b
9. a, c
10. a
11. b

4. b, d
5. c
6. c
7. a
8. b
12. a
13. b, c
14. b
15. c, d
16. a

Completion

1. protonphore, mitochondrial
2. flavoprotein, FeS protein, Q enzyme, cytochrome
3. synthesis, breaking down
4. ADP
5. three, intermembrane space, three
6. Refer to the figure in Section IV of Focused Review
   a. outer membrane
   b. intermembrane space
   c. inner membrane
   d. matrix
   e. proton translocating ATP-ase
   f. cristae
7. flavin, pantothenic acid
8. citric acid, isocitrate, ketone
9. chemiosmotic, the energy in the electron transfer chain creates a proton gradient. There is a higher concentration of $H^+$ in the intermembrane space than inside the mitochondrion. The protons pumped out of the mitochondrion provide the driving force to phosphorylation of ADP by propelling the protons to flow back into the mitochondrion through the proton translocating ATP-ase. This enzyme manufactures the ATP.
10. mitochondrion
11. citrate, CoA
12. $H^+$, electrons
13. oxidizing, reduced, one, 2

"You are what you eat."
- Unknown

# 21

# Specific Catabolic Pathways: Carbohydrate, Lipid, and Protein Metabolism

## CHAPTER OBJECTIVES

After you have studied the chapter and worked the assigned exercises in the text and the study guide, you should be able to do the following.

1. Name the major pathways by which energy is extracted from monosaccharides and fatty acids.
2. Discuss the three stages of glycolysis with regard to the energetics (ATP consumption or generation), the number of carbons in the molecular units and the reaction types involved.
3. Distinguish between anaerobic and aerobic glycolysis.
4. Indicate under what conditions pyruvate is converted to (a) lactate, (b) ethanol or acetyl CoA. Write out the reaction for each process.
5. Tabulate the energy yield produced by reactions involved in the complete metabolism of glucose. Indicate the influence of the glycerol-3-phosphate or the aspartate-malate shuttle in the total energy yield.
6. Indicate the point in the catabolic process at which (a) glycerol and (b) fatty acids converge to a common pathway with monosaccharides.
7. Indicate the types of reactions in the $\beta$-oxidation spiral of fatty acids.
8. Specify the cellular location in which (a) glycolysis, (b) fatty acid activation and oxidation, (c) citric acid cycle, (d) oxidative phosphorylation and the (e) urea acid cycle occur.
9. Compare the energy yields for (a) glucose, (b) glycerol and (c) a fatty acid.

10. Name the ketone bodies and explain why they are found in high concentrations in the blood under certain conditions.

11. Discuss the three stages of nitrogen metabolism.

12. Indicate the relationship between heme, biliverdin, bilirubin, and ururubin.

13. Define the important terms and comparisons in this chapter and give specific examples where appropriate.

## IMPORTANT TERMS AND COMPARISONS

Food
Glycolysis
β-Oxidation
Amino Acid Pool
Oxidative Deamination
Urea Cycle
Fat Deposit
Glucose 6-Phosphate
$C_3$ Fragment
Anaerobic Pathway
Pyruvate
Pyruvate Dehydrogenase
Cytosol
Glycerol-3-Phosphate Transport
Aspartate-Malate Shuttle
Glycerol-1-Phosphate
Fatty Acyl-CoA
Carnitine Acyltransferase
Thiolase
Energy Yield
Ketone Bodies
Urea
α-Ketoglutarate
Carbamoyl Phosphate
Ornithine
Glucogenic and Ketogenic
Amino Acid
Monosodium Glutamate (MGS)
Phenylketonurea (PKU)
Phagocyte
Biliverdin
Bilirubin
Ferritin
Urobilin
Jaundice

## SELF-TEST QUESTIONS

MULTIPLE CHOICE. In the following exercises, select the correct answer from the choices listed. In some cases, two or more answers will be correct.

1. The initial reaction in the β-oxidation spiral of fatty acids produces

a. Acyl CoA
b. Acetyl CoA
c. Carbamoyl phosphate
d. Dihydroxyacetone phosphate

2. The final product of aerobic glycolysis is

a. Pyruvate
b. Acetyl Co
c. Lactate
d. Ethanol

3. The number of ATP molecules produced on the complete metabolism of glucose in skeletal muscle is

a. 24
b. 36
c. 12
d. 38

4. Anaerobic glycolysis occurs in the

a. Mitochondria
b. Cytosol
c. Cytosol and mitochondria
d. Extracellular fluid

5. Ketone bodies are produced in conditions of a

a. High glucose supply
b. Low amino acid pool
c. Low glucose supply
d. Missing fructose-1-phosphate aldolase

6. The heme unit of hemoglobin contains

a. $Fe^{+2}$
b. Globin protein
c. Carbohydrate
d. Glycerol

7. The first reaction involved in nitrogen catabolism of amino acids is a(n)

a. Oxidative Deamination
b. Reduction
c. Hydration
d. Transamination

8. The final step in the urea cycle involves the

a. Splitting of argininosuccinate
b. Oxidation of glutamate
c. Hydrolysis of arginine
d. Transamination of glutamate

9. The order of the caloric value for carbohydrates, proteins and fats is

a. Protein = Fat = Carbohydrate
b. Carbohydrate > Fat > Protein
c. Fat > Carbohydrate > Protein
d. Protein > Fat > Carbohydrate

10. Pyruvate undergoes an oxidative decarboxylation with CoA to produce acetyl CoA and

a. Lactate
b. Glutamate
c. $CO_2$
d. $NAD^+$

11. Which of the following molecules are transported into the mitochondria by the glycerol-3-phosphate transport or the aspartate-malate shuttle?

a. ATP
b. $CO_2$
c. NADH
d. ADP

12. The number of cycles of the β-oxidation spiral that the (saturated) palmitic acid ($C_{16}$) goes through is

a. 8
b. 16
c. 7
d. 4

<u>COMPLETION</u>. Write the word, phrase or number in the blank space in answering the question.

1. The three end products of complete catabolism of proteins in the body are ________, _________, and ______________.

2. Specialized cells that store fats are called ________________________________.

3. In the catabolic metabolism of hemoglobin, the iron is taken up by ___________.

4. ________________ is considered one of the ketone bodies despite not having a ketone functional group.

5. An aldolase enzyme catalyzes the splitting of fructose 1,6-diphosphate to ______________ and ________________.

6. Fats are metabolized by first hydrolyzing them to ________ and _______________.

7. Although ___________ is a basic amino acid and occurs in the urea cycle, it is not found in proteins.

8. Thr β-oxidation of a $C_{10}$ unsaturated fatty acid requires one ___________ step in the process than that for the saturated $C_{10}$ acid.

9. Fatty acid oxidation occurs in the ___________________.

10. If the carbon skeleton of an amino acid is converted to pyruvate, it may serve in either of two roles.

    a. ___________________________

    b. ___________________________

## ANSWERS TO SELF-TEST QUESTIONS

Multiple Choice

1. a
2. a
3. b
4. b
5. c
6. a
7. d
8. c
9. c
10. c
11. c
12. c

Completion

1. urea or $NH_4^+$, $CO_2$, $H_2O$
2. fat deposits
3. ferritin
4. β-hydroxybutyric acid
5. glyceraldehyde 3-phosphate, dihydroxyacetone phosphate
6. glycerol, fatty acid
7. ornithine
8. more
9. mitochondrion
10. a. an energy source
    b. a building block in the synthesis of glucose (and other metabolites)

If we could not store so vital a
substance as glucose in our bodies,
we would have to be eating incessently
to maintain a steady supply of it. That
would be a precarious existence.

-Ernest Borek
The Atoms Within Us

# 22

# Biosynthetic Pathways

## CHAPTER OBJECTIVES

After you have studied the chapter and worked the assigned exercises in the text and the study guide, you should be able to do the following.

1. Contrast anabolic and catabolic reactions in terms of energy requirements and the form of coenzymes involved.
2. Explain why anabolic reaction pathways are not just the reverse of the corresponding catabolic pathways.
3. Using Figure 22.1 (text), identify the major metabolic intermediates which can be used in gluconeogenesis.
4. Discuss the importance of UDP-glucose in carbohydrate biosynthesis.
5. Identify the feature that is common in both fatty acid synthesis and degradation.
6. Describe the role of acetyl CoA and malonyl CoA in fatty acid synthesis.
7. Indicate the form of CoA and the other reactants involved in the synthesis of phosphatidate and sphingosine.
8. Write out the reactions for the synthesis of glutamic acid and the subsequent transamination reaction to form an amino acid.
9. Define the important terms and comparisons in this chapter and give specific examples where appropriate.

## IMPORTANT TERMS AND COMPARISONS

Le Chatelier's Principle
Anabolic Pathway
Glycogen
Photosynthesis
Gluconeogenesis and Glycogenesis
Glucogenic Amino Acid
Uridine Diphosphate (UDP)
Essential Fatty Acid
Synthase
Malonyl ACP
Acyl CoA
Hyaline Membrane Disease

## SELF-TEST QUESTIONS

MULTIPLE CHOICE. In the following exercises, select the correct answer from the choices listed. In some cases, two or more choices will be correct.

1. Gluconeogenesis is associated with the synthesis of:

   a. glycogen
   b. glucogenic amino acids
   c. glucose
   d. glucose-1-phosphate

2. Glycolysis and gluconeogenesis proceed in reverse directions. However, the number of points in this pathway in which there are enzymes not common to each pathway is:

   a. 6
   b. 8
   c. 4
   d. 2

3. Both fatty acid synthesis and degradation have the common feature of involving:

   a. ATP
   b. $NAD^+$
   c. acetyl CoA
   d. ketogenic amino acids

4. In each cycle in fatty acid synthesis, the unit added on is a:

   a. $C_1$ fragment
   b. $C_2$ fragment
   c. $C_4$ fragment
   d. $C_6$ fragment

5. Which of the following structures represent UDP-glucose?

a. 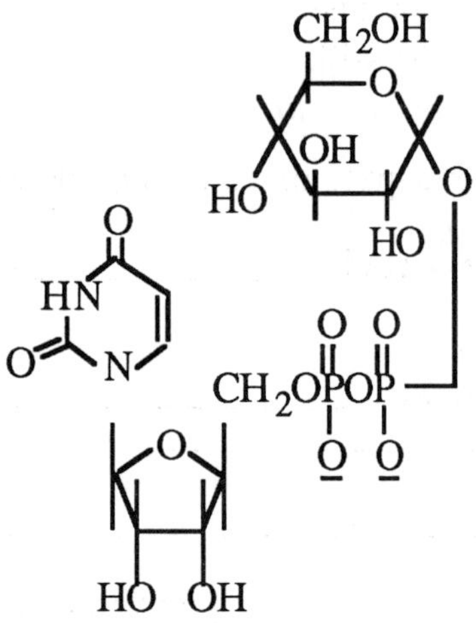

c. 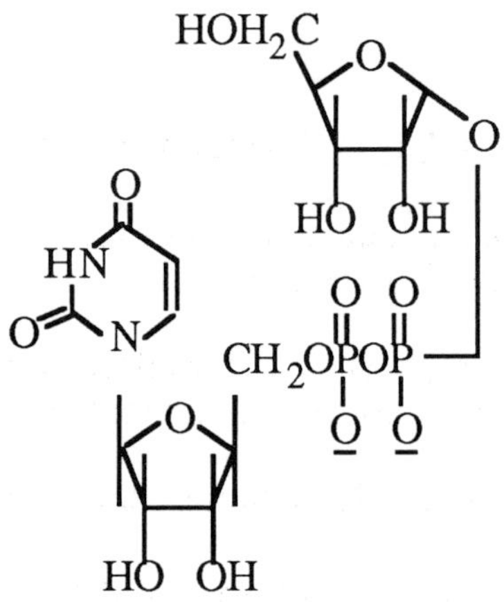

b. 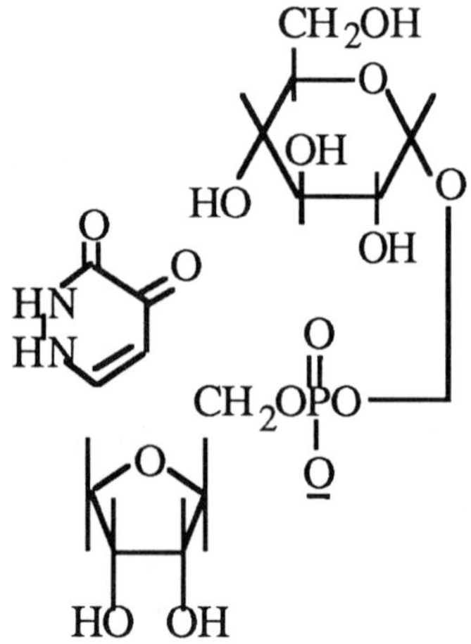

d. 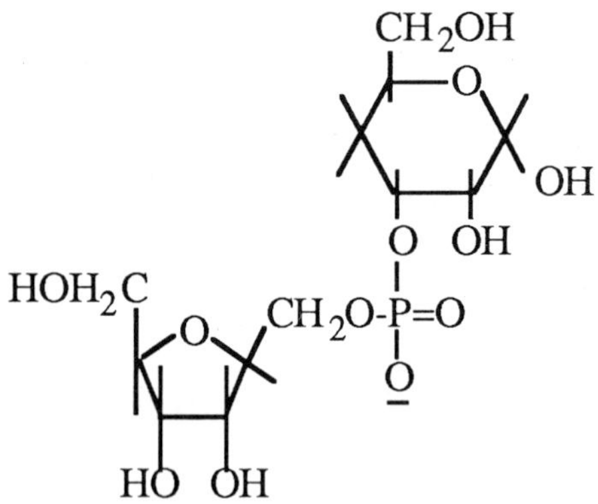

6. The amino acid produced in the following reaction is:

$$\begin{array}{c} CH_3 \\ | \\ C{=}O \\ | \\ CO_2^- \end{array} \quad + \quad \begin{array}{c} CO_2^- \\ | \\ HCNH_3^+ \\ | \\ CH_2 \\ | \\ CH_2 \\ | \\ CO_2^- \end{array} \quad \begin{array}{c} \text{------>} \\ \text{<-----} \end{array}$$

a. glycine
b. glutamic acid
c. arginine
d. alanine

7. Which of the following can be used in gluconeogenesis?

a. pyruvate
b. lactate
c. ketogenic amino acids
d. urea

8. Which of the following are essential fatty acids?

a. stearic acid
b. butyric acid
c. palmitic acid
d. linoleic acid

COMPLETION. Write the word, phrase or number in the blank space in answering the question.

1. Biosynthetic pathways require an ___________ input and require _____________ power in the form of NADPH.

2. Glycogen is synthesized from the activated glucose monomer, _________________, which is formed in the reaction of __________ and _________________.

3. The overall reaction in photosynthesis is _____________________________________.

4. Fatty acid synthesis occurs in the _____________ , while fatty acid degradation takes place in the ____________ .

5. The acetyl or malonyl group in the ACP complex is attached through a __________ atom.

6. The fatty acid component of phosphatidate is derived from the ____________ which reacts with glycerol-3-phosphate.

7. The biosynthesis of cholesterol occurs in the ____________ (an organ) and is assembled from _______________.

8. The phosphatidylcholine in normal lungs contains predominantly ________________ fatty acids.

9. Photosynthesis occurs in the _________ organelles of plants.
The first reaction in photosynthesis requires energy (sunlight) and is called the _________ reaction. The four important species generated are ________ , ________ , ___________ and __________ .
The second reaction uses $CO_2$ and three of the species generated in the light reaction , ________ , ___________ and ________ to produce _____________ macromolecules.

## ANSWERS TO SELF-TEST QUESTIONS

Multiple Choice

1. c
2. c
3. c
4. b
5. a
6. d
7. a, b
8. d

Completion

1. energy, reducing
2. UDP-glucose, UTP, glucose-1-phosphate
3. $nH_2O + nCO_2 \xrightarrow{\text{sunlight}} (CH_2O)_n + nO_2$
4. cytosol, mitochondria
5. S
6. Acyl CoA
7. liver, acetyl CoA

8. unsaturated
9. chloroplasts;
   light, ATP, NADPH, $H^+$, $O_2$;
   ATP, NADPH, $H^+$, carbohydrate

What distinguishes a butterfly from a lion, a hen from a fly, or a worm from a whale is........

-F. Jacob
Science, 196, 1161 (1977)

# 23

# Nucleic Acids and Protein Synthesis

## CHAPTER OBJECTIVES

Each one of us is unique in many ways. These characteristics result because each individual has a different genetic makeup in that, although our DNA appears much the same as everyone else's, the sequence (i.e., the order of the bases) of nucleotides in the DNA strands is quite different. These differences become manifest as the genes in the DNA are expressed into different types and amounts of proteins which reveal our individuality. In this chapter, we learn about the two types of nucleic acids - DNA and RNA. After first realizing the constituents and structure of these polynucleotide molecules, we proceed into defining their essential functions as they relate to the Central Dogma of Molecular Genetics. The "flow" of genetic information is briefly summarized on the following page.

| Genetic Material | | Genetic Message | | Genetic Protein Product |
|---|---|---|---|---|
| DNA | Transcription (RNA Synthesis) ⟹ | mRNA, tRNA, rRNA | Translation (Protein Synthesis) ---> | Protein |

Replication

(DNA Synthesis)

into the

a. Amino acid specific tRNAs bring in the "activated form" of the amino acid. Correct base pairing between the anti-codon triplet (tRNA) and the codon (mRNA) occurs. The amino acid which becomes incorporated protein is defined by the genetic code.

b. rRNA combines with proteins to form ribosomes, on which translation occurs.

Nucleus | Cytoplasm or Cytosol

The complexity of DNA is further revealed in the exon and intron segments of the genes in higher organisms and the repeating nucleosome subunits in the nuclear chromosome. These properties were only realized in the mid 1970's. The delicate nature of DNA is brought out by the potentially damaging effect of an error or mutation in a single nucleotide in DNA. Finally, the chapter concludes with the general aspects of and the potential uses of recombinant DNA technology.

After you have studied the chapter and worked the assigned exercises in the text and the study guide, you should be able to do the following.

1. Explain the importance of hereditary information and its storage and expression in DNA.

2. Name and draw the structures of the organic bases (purines and pyrimidines), nucleosides and nucleotides that occur in DNA and RNA.

3. Draw a dinucleotide unit of DNA containing one purine and one pyrimidine base; point out the 5'- and 3'-ends, the glycosidic bond between the sugar and base and the phosphodiester backbone in this unit.

4. Draw the complementary base pairs, G≡C, and A=T, showing all atoms and the hydrogen bonding interactions.

5. In double-stranded DNA, indicate the relative positions of the aromatic bases, the sugar and phosphate units in this structure.

6. List the compositional, structural and functional differences between DNA and RNA.

7. Outline the main features of DNA replication, indicating the role or significance of unwinding proteins, DNA polymerase, complementary base pairing, semi-conservative replication, 5'----->3' direction, Okazaki fragments and DNA ligase.

8. Outline the transcription process and the role of mRNA and the codon units in the transferring of genetic information.

9. Describe the structure of a tRNA molecule, the position of the anticodon loop, the role of the 3'-end and the function of this RNA in the translation process.

10. Describe the location and role of rRNA in the translation process.

11. Explain the importance of the genetic code.

12. Describe the processes of amino acid activation and then the initiation, elongation and termination steps in protein synthesis.

13. Define the term gene as it relates to higher organisms.

14. Describe the nucleosome unit in terms of its composition and structure.

15. Explain how a mutation can be produced during DNA replication or by chemicals called mutagens.

16. Explain the use of plasmids, restriction endonucleases and Eschericia coli in recombinant DNA technology.

17. Define the important terms and comparisons in this chapter and give specific examples where appropriate.

## IMPORTANT TERMS AND COMPARISONS

Zygote
Nucleus (of the cell)
Histone Protein
Nucleic Acid
Deoxyribonucleic Acid (DNA)
One Gene-One Enzyme Hypothesis
Ribonucleic Acid (RNA)
Purine Bases; A, G
Pyrimidine Bases; C, T, U
Anticancer Drug
D-Ribose and D-Deoxyribose
Nucleoside and Nucleotide
Glycosidic Linkage
3'- and 5'-Carbon on Sugar
Primary Structure
3'-and 5'-OH End of DNA
Complementary Base Pairs
A=T and G≡C Hydrogen Bonding
DNA Polymerase
Semiconservative Replication
Daughter Strand
Leading and Lagging Strand
DNA Ligase
Clover Leaf Structure
Central Dogma of Molecular Biology
Transcription and Translation
RNA Polymerase
Ribonucleotide and Deoxyribonucleotide
Antibiotic
Svedberg Unit, S
Codon in mRNA
Anticodon in tRNA
Cytosol
Activated Amino Acid Synthetase
Amino Acid-tRNA
40 S and 60 S Ribosomes
P and S Site
Transferase
(Inheritable) Gene
Exon and Intron
Virus
Mutagen and Mutation
Recessive and Dominant Gene
Eschericia coli
Plasmid DNA
Xeroderma Pigmentosum
"Cut, Patch, Cut and Seal" Process
Splice

## SELF-TEST QUESTIONS

MULTIPLE CHOICE. In the following exercises, select the correct answer from the choices listed. In some cases, two or more choices will be correct.

1. The backbone in a nucleic acid is called a:

   a. glycosidic bond
   b. sugar backbone
   c. phosphodiester backbone
   d. peptide backbone

2. The coding sequences of a gene in DNA are called:

   a. nucleosomes
   b. introns
   c. exons
   d. initiating factors

3. An RNA virus contains which of the following:

   a. DNA
   b. RNA
   c. proteins
   d. carbohydrates

4. A nucleosome is made up of:

   a. DNA and histone proteins
   b. rRNA and proteins
   c. nucleosides
   d. both DNA and RNA

5. The Ames test is used to:

a. verify the genetic code
b. study replication
c. distinguish tRNA from mRNA
d. test for mutagens

6. A feature which distinguishes ribose from deoxyribose is the absence of an OH at the:

a. 2'-carbon
b. 3'-carbon
c. 1'-carbon
d. 5'-carbon

7. In protein synthesis, the initial amino acid is always:

a. glycine
b. cysteine
c. methionine
d. alanine

8. Adenine is the name of a:

a. nucleoside
b. purine base
c. nucleotide
d. amino acid

9. The sequence GCCCTGA has the A at the:

a. 5'-end
b. 3'-end
c. can be either the 5'- or 3'- end
d. depends on size of DNA

10. The 3'-end of a tRNA molecule:

a. contains the anticodon loop
b. binds to a specific amino acid
c. has no known function
d. binds to rRNA

11. The structure of cytosine is:

a. 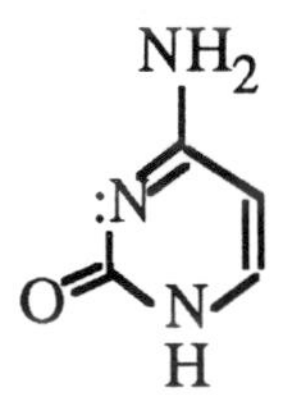

b. 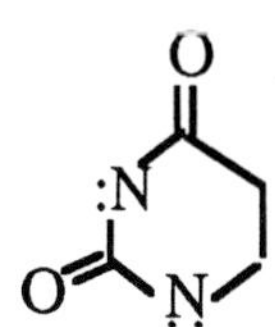

c. 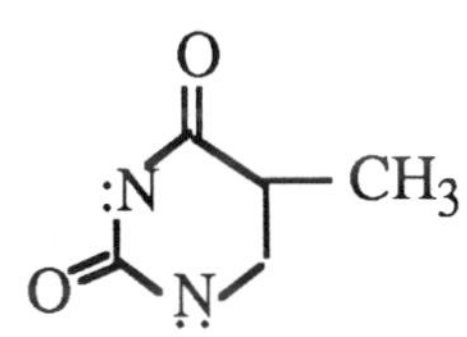

d. 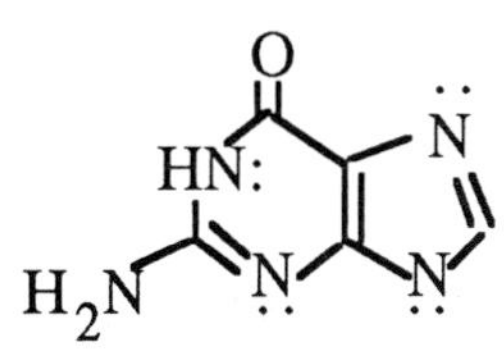

12. The initiating codon in protein synthesis is:

a. GGG
b. GUA
c. AUG
d. ATG

13. Antibiotics are used to:

a. kill viruses
b. inhibit DNA replication
c. bind to mutagens
d. kill bacteria

14. Okazaki fragments are:

a. formed in RNA transcription
b. formed in DNA replication
c. small RNA molecules
d. small DNA fragments of about 1000 nucleotides each

15. The glycosidic bond is between the ribose sugar and which position on a purine base?

a. N-1 atom
b. N-9 atom
c. N-3 atom
d. C-8 atom

16. Complementary base pairs in DNA are:

a. AU
b. GC
c. AT
d. GT

17. In protein synthesis, the mRNA attaches to the:

a. 30 S ribosome
b. 40 S ribosome
c. 100 S nucleosome
d. 60 S ribosome

18. An RNA molecule has no:

a. uracil bases
b. thymine bases
c. cytosine bases
d. guanine bases

19. The hydrogen bonding in an AT base pair involves atoms in the purine ring associated with the:

a. 6-membered ring
b. 5-membered ring
c. both the 5- and 6-membered ring
d. none of the above

20. An amino acid is activated by reaction with:

a. tRNA
b. synthetase
c. ATP
d. methionine

21. The antibiotic, streptomycin, kills bacteria by stopping:

a. replication
b. transcription
c. formation of Okazaki fragments
d. translation

22. The process by which cells become specialized (neurons, muscle,etc.) in the development of a multicellular organism is called:

a. replication
b. differentiation
c. cell division
d. genetic mutation

23. The operon is a segment of DNA which collectively includes:

a. collection of structural genes
b. control sites
c. regulatory genes
d. all of the above

24. Methylation of DNA occurs on which of the following bases:

a. adenine
b. guanine
c. thymine
d. cytosine

COMPLETION. Write the word, phrase or number in the blank space in answering the question.

1. A DNA molecule may have as many as a ____________ genes.

2. Replication of DNA occurs in the __________ of the cell, while translation occurs in the __________________.

3. There are __________ hydrogen bonds between adenosine and the complementary base ____________________.

4. Human insulin is one of the first proteins to be produced by ____________ ____________ _________________.

5. A __________ is produced on the joining of an egg cell with a sperm cell.

6. All amino acids are attached to the _____________end of tRNA on the _________________ unit.

7. Referring to the table for the Genetic Code in the text, write the amino acid sequence for the protein produced from the mRNA shown below:

   AUGUUGCACAACGGGGCGGUGGUGUAA

8. After combining the DNA of a plasmid that has sticky ends with another fragment of DNA with the same sticky ends, the enzyme ___________________ ____________________ is added to covalently splice the two DNA fragments together.

9. The organic bases in DNA reside on the interior of the double helix and stabilize the structure by _____________ interactions.

10. _____________ bases are made up of two fused rings, while ________________ bases have only a six-membered heterocyclic ring.

11. Ionizing radiation, gamma rays and chemicals, including benzene and vinyl chloride, are known to be ___________________.

12. The sequence of one strand of double stranded DNA is (5')AGGAGCTTCG(3'). The sequence of the complementary strand is _______________________.

13. Synthesis of RNA (transcription) occurs in a ________________ direction and along the complementary DNA strand in a ________________ direction.

14. Match the scientist(s) in the left-hand column with the accomplishment on the right with which he (they) is most closely associated.

| | | | |
|---|---|---|---|
| a. | B. Ames | (i) | Postulated structure of DNA |
| b. | J. Watson & F. Crick | (ii) | Proposed DNA contains the genetic information |
| c. | A. Klug | (iii) | X-ray studies of DNA |
| d. | O. Avery | (iv) | Proposed one gene-one enzyme hypothesis |
| e. | R. Franklin & M. Wilkins | (v) | Discovered that in DNA, no. of A = no. of T no. of G = no. of C |
| f. | G. Beadle & E. Tatum | (vi) | Elucidated the structure of the nucleosome |
| g. | M. Nirenberg | (vii) | Experimentally revealed the genetic code |
| h. | E. Chargaff | (viii) | Developed test for mutagens |

## ANSWERS TO SELF-TEST QUESTIONS

Multiple Choice

1. c
2. c
3. b, c
4. a
5. d
6. a
7. c
8. b
9. b
10. b
11. a
12. c
13. d
14. b, d
15. b
16. b, c
17. b
18. b
19. a
20. c
21. d.
22. b
23. d
24. d

Completion

1. million
2. nucleus; cytoplasm (or cytosol)
3. two; thymine
4. recombinant DNA technology
5. zygote
6. 3'; adenosine
7. (N-terminal) Met-Leu-His-Asn-Gly-Ala-Val-Val (C-terminal)
8. DNA ligase
9. hydrophobic

10. purine; pyrimidine
11. mutagens
12. (3')TCCTCGAAGC(5')
13. 5'------>3'; 3'------>5'
14. a. (viii)
    b. (i)
    c. (vi)
    d. (ii)
    e. (iii)
    f. (iv)
    g. (vii)
    h. (v)

One, if by land, or two, if by sea.
- Paul Revere's Ride
Henry Wadsworth Longfellow

# 24

# Chemical Communication: Neurotransmitters, Hormones and Immunoglobulins

## CHAPTER OBJECTIVES

After you have studied the chapter and worked the assigned exercises in the text and the study guide, you should be able to do the following.

1. Tabulate the three general types of communication chemicals, indicating (1) their general make up, (2) what each acts on, (3) the time frame of their action, (4) the cells or organs which are involved in the communication and (5) whether secondary messengers are involved in the process.

2. Explain the role of the axon, synapse, vesicle, receptor sites and dendrites in the neurotransmission of acetylcholine.

3. Describe how acetylcholine bound to postsynaptic receptor sites can be removed normally in the cell and how muscle relaxants or nerve gases can prevent this.

4. Outline the mechanism of action and the final inactivation of adrenergic neurotransmitters.

5. List the three types of hormones, their general make up and the types of action that each stimulates.

6. Draw the general structure of an antibody and indicate the types of interactions which occur with specific antigens to produce a sequestered foreign material.

7. Define the important terms and comparisons in this chapter and give specific examples where appropriate.

## IMPORTANT TERMS AND COMPARISONS

Mediator and Receptor
Lymphocyte
Axon and Dendrite
Synapse, Presynapse and Postsynapse
Vesicle
Cholinergic, Monoamine, Adrenergic
    and Peptidergic Neurotransmitter
Acetylcholine
Acetylcholinesterase
Alzheimer's Dementia
Nerve Gas
Cyclic AMP
Primary Messenger
Secondary Messenger
AIDS, T lymphocytes and AZT
Adenylate Cyclase
Phosphodiesterase
L-DOPA
Histamine Receptors $\mathbf{H_1}$ and $\mathbf{H_2}$
GABA
Agonist and Antagonist
Enkephalin
Diabetes
Antigen and Antibody
Phagocytosis
Immune Response
Precipitin Reaction
Vaccine
Lymphocyte Memory Cell
Lymphocyte Plasma Cell

## SELF-TEST QUESTIONS

MULTIPLE CHOICE. In the following exercises, select the correct answer from the choices listed.

1. Decamethionium, which has a structure similar to that for acetylcholine, inhibits acetylcholinesterase. It is a

   a. non-competitive inhibitor
   b. chemical mediator
   c. competitive inhibitor
   d. toxic poison

2. Of the three types of communication chemicals presented in the chapter, the one which acts the slowest is

   a. immunoglobulins
   b. hormones
   c. neurotransmitters
   d. GABA

3. Acetylcholinesterase hydrolyzes which linkage in acetylcholine, $H_3C\underset{\underset{O}{\|}}{C}OCH_2CH_2\overset{+}{N}(CH_3)_3$.

   a. $H_3C\text{----}\overset{\downarrow}{\phantom{-}}\underset{\underset{O}{\|}}{C}$ (arrow at the C----C bond)
   b. $H_3C\underset{\underset{O}{\|}}{C}\text{----}O$ (arrow at the C----O bond)
   c. $H_3C\underset{\underset{O}{\|}}{C}O\text{----}C$ (arrow at the O----C bond)
   d. $H_3\underset{|}{C}\text{----}N(CH_3)_3$ (arrow at the C----N bond)

4. $H_1$ receptors found in the respiratory tract are blocked by the interaction of

a. insulin
b. enkephalins
b. L-DOPA
d. antihistamines

5. The second messenger in the action of monoamine neurotransmitters is in most cases

a. d-AMP
c. AMP
b. c-AMP
d. ATP

6. The administration of L-DOPA can reverse the symptoms of Parkinson's disease. Dopamine is produced from L-DOPA in a

a. hydrolysis reaction
b. oxidative decarboxylation reaction
c. decarboxylation reaction
d. oxidative reaction

7. Protein kinases function as

a. phosphate transfer enzymes
b. structural membrane proteins
c. secondary messengers
d. ion translocating proteins

8. Neurotransmitters are stored in vesicles in the

a. axon
c. presynaptic site
b. dendrites
d. postsynaptic site

9. Glucagon is an example of a

a. peptide hormone
b. steroid hormone
c. monoamine neurotransmitter
d. receptor

10. An agonist is a molecule which

a. stimulates antibody production
b. competitively inhibits acetylcholinesterase
c. activates receptor sites
d. reduces pain

11. Steroid hormones produce their effect in the cell by

a. binding to the membrane
b. activating a protein kinase
c. stimulating c-AMP
d. affecting the transcription of genes

COMPLETION. Write the word, phrase or number in the blank space in answering the question.

1. Complete the following reactions, writing structural formulas for products and/or reactants.

a. $(CH_3)_3\overset{+}{N}CH_2CH_2O\overset{\overset{\displaystyle O}{\|}}{C}CH_3 + H_2O$ -------> ____________ + ____________

b. 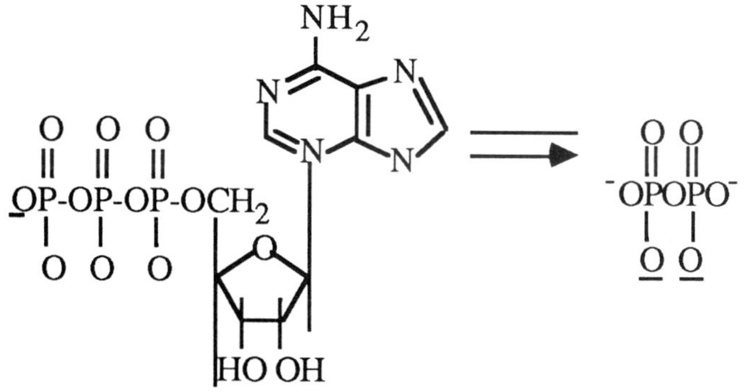

____________________

c. ____________ + $HOCH_2CH_2\overset{+}{N}(CH_3)_3$ -------> ________ + _________

2. The ____________ is the fluid-filled space between the axon of one neuron and the ______________ of the next neuron.

3. A neurotransmitter and a receptor combine very specifically together. This interaction has been likened to the _____________ __________ _____________ model for enzyme-substrate interactions.

4. Food poisoning from botulinus toxin deprives cholinergic nerves from their transmitters. This indicates that the synthesis of _____________ is blocked.

5. To degrade the neurotransmitter ,acetylcholine, in the postsynaptic receptor site, the enzyme, ______________ , acts on acetylcholine to ______________ it to acetic acid and choline.

6. Monoamine neurotransmitters, such as norepinephrine, form a membrane bound ____________ - _______________ like complex which activates the enzyme __________which produces the secondary messenger ______________ from ATP.

7. GABA is the abbreviation for ____________________________ .

8. The ________________ end of methionine enkephalin in similar to a segment of the alkaloid painkiller, ________________.

9. Antidiabetic drugs are useful for diabetic patients who have insufficient numbers of ______________ ________________ on their target cells.

10. An immunoglobulin molecule is a ______________ having ______________ polypeptide chains covalently linked together with ________________ bonds. The ______________ -terminal ends of all polypeptide chains contain a variable amino acid region, while the ______________ -terminal ends contain the segments referred to as the constant regions.

11. The AIDS virus primarily invades and kills ______ lymphocyte cells. The most effective therapeutic drug used today in the treatment of AIDS is __________ .

## ANSWERS TO SELF-TEST QUESTIONS

Multiple Choice

1. c
2. a
3. c
4. d
5. b
6. c
7. a
8. c
9. a
10. c
11. d

Completion

1. a. acetylcholinesterase; $H_3CCO_2H$ + choline
   b. adenylate cyclase; c-AMP (see figure in text)
   c. $H_3C\overset{\overset{O}{\|}}{C}\text{-}SCoA$; acetylcholine transferase; $H_3C\underset{\underset{O}{\|}}{C}OCH_2CH_2\overset{+}{N}(CH_3)_3$ + CoA-SH

2. synapse; dendrites (or cell body)
3. lock-and-key
4. acetylcholine
5. acetylcholinesterase; hydrolyze
6. hormone-receptor; adenylate cyclase; c-AMP
7. $\gamma$-aminobutyric acid
8. N-terminal end; morphine
9. insulin receptors
10. glycoprotein; 4; (S-S) or disulfide; N or amino; C or carboxyl
11. T, AZT (azidothymidine)

"The thing to remember when traveling is that the trail is the thing, not the end of the trail. Travel too fast and you miss all you are traveling for."
- Louis L'Amour

# 25

# Body Fluids

## CHAPTER OBJECTIVES

After you have studied the chapter and worked the assigned exercises in the text and the study guide, you should be able to do the following.

1. Tabulate the functions of blood and indicate which molecules play a major role in these functions.
2. Characterize each of the three main blood cell types, including where they are formed and their major functions.
3. Compare the components of blood, blood plasma, and serum.
4. Discuss the interrelationship between prothrombin, thrombin, thromboplastin, and thrombosis.
5. Outline the characteristics of the hemoglobin molecule, its $O_2$ binding characteristics and how the $CO_2$ and $H^+$ influence $O_2$ binding.
6. Explain the role of nephrons in retaining proteins and vital nutrients, while eliminating waste products from the blood.
7. Discuss the role of the hormones vasopressin and aldosterone in the control of water and salt balance in the blood and kidneys.
8. Define the important terms and comparisons in this chapter and give specific examples where appropriate.

## IMPORTANT TERMS AND COMPARISONS

Extracellular, Interstitial and Intercellular Fluid
Erythrocytes, Leukocytes and Platelets
Plasma and Serum
Clotting
Fibrinogen and Fibrin
Prothrombin and Thrombin
Thromboplastin
Hemoglobin
Oxygen Dissociation Diagram
Cooperative Action
Synthetic Blood Substitute
Carbaminohemoglobin
Carbonic Anhydrase
LeChatelier's Principle
Acidosis
Alkalosis
Bowman's Capsule
Tubule
Proximal Tubule, Henle's Loop and Distal Tubule
Urine
Vasopressin
Arterial and Venous Capillaries
Rennin
Angiotensinogen and Angiotensin
Tachycardia

## SELF-TEST QUESTIONS

MULTIPLE CHOICE. In the following exercises, select the correct answer from the choices listed. In some cases, two or more choices will be correct.

1. The plasma contains

   a. albumin
   b. erthrocytes
   c. hemoglobin
   d. globulin

2. The synthetic blood substitute containing perfluorocarbon

   a. dissolves $O_2$
   b. dissolves $CO_2$
   c. contains antigens
   d. contains clotting factors

3. The Bohr effect is associated with the

   a. blood filtration in the kidneys
   b. action of hormones in the blood
   c. salt balance in the cell
   d. pH effect on $O_2$ binding to hemoglobin

4. Blood clotting can be prevented during transfusions by the addition of sodium citrate. This agent

   a. complexes with $Ca^{+2}$
   b. converts thrombin back to prothrombin
   c. produces edema
   d. degrades collagen

5. Extracellular fluids are those

   a. within the cell
   b. not inside the cell
   c. found in cytosol
   d. with little or no $Cl^-$

6. A molecule of $CO_2$ binds to hemoglobin to form carbaminohemoglobin. The $CO_2$ forms a covalent bond at the

   a. Fe(II) in the heme group
   b. amino-terminal end of the polypeptide chain
   c. carboxyl-terminal end of the polypeptide chain
   d. $H^+$ binding site

7. Carbonic anhydrase catalyzes which of the following reactions.

   a. $Hb + O_2 \rightleftharpoons HbO_2$      Hb = hemoglobin

   b. $CO_2 + H_2O = H_2CO_3$

   c. $HbO_2 + H^+ + CO_2 = Hb(CO_2)H^+ + O_2$

   d. $H_2PO_4^- = H^+ + HPO_4^{-2}$

8. Vasopressin is

   a. a diuretic hormone
   b. called ADH
   c. a small peptide
   d. manufactured in the hypophysis

9. Hemoglobin more readily binds to the second, third and fourth $O_2$ molecules than it does the first because of

   a. the Bohr effect
   b. an allosteric effect
   c. a pH effect
   d. Henle's loop

10. Some of the functions of the blood include

   a. maintaining the pH of the body
   b. fighting infection
   c. carrying nutrients
   d. carrying $O_2$ from the lungs to the tissues

11. The cells that are made in the bone marrow include

   a. erythrocytes
   b. leukocytes
   c. platelets
   d. nephrons

12. The serum does not contain

   a. albumin
   b. globulins
   c. fibrinogen
   d. water

<u>COMPLETION</u>. Write the word, phrase or number in the blank space in answering the question.

1. Severe dehydration stimulates the production of the hormone _______________ to enable the body to retain more _______________.

2. Hormonal control of blood pressure involves the secretion of the enzyme _______________ from the kidney. This enzyme converts the inactive protein, _______________, into the potent vasoconstrictor, _______________.

3. The serum contains all the elements of plasma except for ________________.

4. Persons with hypertension have ______________ blood pressure. This can lead to a number of medical problems, including ______________ or ________________.

5. The ______________ molecule is bound to the ______________ in the fully oxygenated form of hemoglobin.

6. ________________ is the major component of the blood which is a liquid and contains no ____________.

7. ____________ is an ______________ which initiates blood clotting. It is ordinarily found in the blood in the inactive form called ______________.

8. Most of the $CO_2$, a waste product in cellular metabolism, is transported to the lungs as ______________.

9. The ____________ organ acts a superfiltration unit, in which the waste products are eliminated in the ____________.

10. ______________ is the name of the condition which occurs when the pH of the blood increases above normal.

11. A nephron contains ____________ capsule which is connected to a tiny tube called the ____________. The U-shaped twist in the unit is called ______________ ______________.

ANSWERS TO SELF-TEST QUESTIONS

Multiple Choice

1. a, d
2. a, b
3. d
4. a
5. b
6. b
7. b
8. b, c, d
9. b
10. a, b, c, d
11. a, b, c
12. c

Completion

1. aldosterone; water
2. rennin; angiotensinogen; angiotensin
3. fibrinogen
4. high; heart attack; stroke; kidney failure (any two of these)
5. $O_2$; Fe in the heme group
6. plasma; cells
7. thrombin; enzyme; prothrombin
8. $H_2CO_3$
9. kidney, urine
10. alkalosis
11. Bowman's; tubule; Henle's loop

"I hate quotations. Tell me what you know."
- Ralph Waldo Emerson

# Nutrition and Digestion

## CHAPTER OBJECTIVES

After you have studied the chapter and worked the assigned exercises in the text and the study guide, you should be able to do the following.

1. Define nutrient and classify the nutrients into the six groups.
2. Indicate what RDA stands for.
3. Explain the difference between basal caloric requirement and the normal caloric intake of active individuals.
4. Explain why a lengthy diet is usually necessary to lower weight by more than a few pounds.
5. Indicate the primary roles in the body for carbohydrates, fats and proteins.
6. Distinguish between digestible and non-digestible carbohydrates.
7. Distinguish between a complete and an incomplete protein from a nutritional standpoint.
8. Define generally the process of digestion.
9. Outline the digestion process for carbohydrates and point out the role of the most important enzymes involved.
10. Outline the digestion process for lipids and point out the role of the most important enzymes involved.
11. Outline the digestion process for proteins and point out the role of the most important enzymes involved.
12. Define the important terms and comparisons in this chapter and give specific examples where appropriate.

## IMPORTANT TERMS AND COMPARISONS

Nutritional Calorie
Essential Fatty Acid
Kwashiorkor
Acetal Linkage
Hydrolysis
α- and β-Amylase
Blood Sugar
Diabetis Mellitus
Cellulase
Bile Salts
Pepsin
Chymotrypsin
Trypsin
Carboxypeptidase

## SELF-TEST QUESTIONS

MULTIPLE CHOICE. In the following exercises, select the correct answer from the choices listed. More than one correct answer may be possible in these questions.

1. Which of the following are classified as nutrients?

   a. Carbohydrates
   b. Nucleic Acids
   c. Esters
   d. Vitamins
   e. Water

2. A nutrient calorie (Cal.) is equivalent to:

   a. 100 calories
   b. 1 calorie
   c. 1000 calories
   d. no fixed number because it will depend on the route of digestion

3. The most prevalent form of dietary fats is:

   a. Fatty Acids
   b. Cholesterol
   c. Triglycerides
   d. Sphingolipids

4. The essential fatty acids are:

   a. Linolenic and Linoleic Acids
   b. Acetic and Linoleic Acids
   c. Palmitic and Linolenic Acids
   d. Stearic and Linoleic Acids

5. Water makes up what percent of body weight:

   a. 15%
   b. 75%
   c. 35%
   d. 60%

6. A number of enzymes are necessary for the digestion of carbohydrates. The role of β-amylase is to:

   a. Hydrolyze the α (1-4) glycosidic linkages in an orderly fashion, cutting disaccharide maltose units one by one from the nonreducing end of the chain
   b. Randomly cut the α (1-4) glycosidic linkages
   c. Hydrolyze the α (1-6) glycosidic linkage
   d. Hydrolyze all glycosidic linkages, irrespect of the type of linkage

7. A carbohydrate that cannot be hydrolyzed in our digestive system is:

   a. Glycogen
   b. Cellulose
   c. Amylose
   d. Amylopectin

8. Some of the final digestion products of lipids as a result of lipase action are:

   a. Glycerol
   b. Choline
   c. Phospholipids
   d. Sphingosine

9. Digestion of proteins results from the hydrolysis of the :

   a. Phosphodiester Linkage
   b. Peptide Linkage
   c. Glycosidic Backbone
   d. Hemiacetal Linkage

10. Enzymes involved in the digestion of proteins include:

   a. Pepsin
   b. Lipase
   c. Carboxypeptidase
   d. α-Amylase

<u>COMPLETION</u>. Write the word, phrase or number in the blank space in answering the following questions.

1. In order to lose 10 pounds of fat, a person must use up __________ Cals.

2. We must consume certain amounts of vitamins, minerals and nutrients. Typically, we must take in __________ of nutrients, while we need only __________ of vitamins and minerals.

3. Most of the minerals are metal ions. Five of the metal ions are ______________, ______________, ______________, ______________, and ______________.

4. Two minerals which are not metals are ______________ and ______________.

5. Two fat soluble vitamins are ______________ and ______________.

6. Four water soluble vitamins are ______________, ______________, ____________ and ______________.

7. Linolenic and linoleic are essential fatty acids in higher animals. They both contain ______________ carbons in the make-up of the fatty acid.

8. Three water soluble vitamins which act as coenzymes in biological reactions are ______________, _______________ and ________________.

9. ________________ is a complete protein, while examples of proteins which are not complete are ______________ and _______________.

10. ______________ are the most concentrated form of energy that we consume as food.

## ANSWERS TO SELF-TEST QUESTIONS

### Multiple Choice

1. a, d, e
2. c
3. c
4. a
5. d
6. a
7. b
8. a, b, d
9. b
10. a, c

### Completion

1. 35,000 Cal.
2. (many) grams; micro- or milligrams
3. Any five of Na, Ca, Mg, Fe, Zn, Cu, Mn, Cr, Mo or Cu
4. Any two of Cl, P, Se, I or F.
5. Any two of vitamins A, D, E or K.
6. Any four of vitamins B1, B2, B6, B12, niacin, folic acid, pantothenic acid, biotin, or vitamin C.
7. 18
8. Any 3 of vitamins B1, B2, B6, B12, niacin or folic acid.
9. Casein; Any two of gelatin or corn, rice or wheat protein.
10. Fats

ANSWERS TO EVEN-NUMBERED

END-OF-CHAPTER PROBLEMS

to accompany

INTRODUCTION TO GENERAL, ORGANIC AND BIOCHEMISTRY

by

F. Bettelheim and J. March

## PREFACE

This section of the study guide provides the answers to the even-numbered, end-of-chapter problems found in the text, Introduction to General, Organic and Biochemistry by Bettelheim and March. This compilation, taken together with the answers cataloged at the end of the text, provides the answers to all the problems at the end of each chapter.

In problem solving, it is not only reassuring to know that the correct answer has been obtained, but at times, it may also provide the needed reinforcement at a pivotal point in your understanding. It is with this philosophy that this answer supplement is included in the manual.

William M. Scovell
Department of Chemistry
Bowling Green State University
Bowling Green, Ohio 43403

# *ANSWERS TO END-OF-CHAPTER PROBLEMS*

## *Chapter 1. Matter, Energy, and Measurement*

1.12 See Section 1.1.
1.14 We have more confidence in a theory than in a hypothesis because the former is better supported by the evidence.
1.16 (a) gram (b) meter (c) liter
1.18 1000 g
1.20 See Section 1.3.
1.22 (a) 98.6°F, 310 K (b) 32°F, 273 K
(c) 482°F, 523 K (d) -58°F, 223 K
1.24 (a) 0.711 kg (b) 45 700 g (c) 73 000 mg (d) 83 400 mm
(e) 4.75 m (f) 63.7 cc (g) 0.0964 L (h) 3290 cc
1.26 44 miles/hour
1.28 See Section 1.6.
1.30 3.3 mL
1.32 221 g
1.34 7.14 g/cc
1.36 See Section 1.8.
1.38 (a) 2100 cal (b) 5100 cal (c) 99 000 cal (d) 2500 cal
1.40 Extensive: (a)(b)(c)(e)(i) Intensive: (d)(f)(g)(h)
1.42 It was a wrong hypothesis. The child did not have lead poisoning.
1.44 190 mg
1.46 630 mg
1.48 normal urine
1.50 Before laser treatments it was necessary to perform an operation to remove kidney stones. Now they can often be removed without surgery.
1.52 to generate heat by muscle action
1.54 (a) $7.426 \times 10^3$ (b) $3.4 \times 10^7$ (c) $6.64 \times 10^{-2}$
(d) $3.62 \times 10^{-9}$
1.56 (a) $6.48 \times 10^7$ (b) $4.69 \times 10^5$ (c) $1.6 \times 10^5$
(d) $2.8 \times 10^{-15}$
1.58 (a) $9.3 \times 10^6$ (b) $6.91 \times 10^5$ (c) $3.733 \times 10^{-4}$
1.60 $1.9 \times 10^4$
1.62 (a) 3 (b) 2 (c) 3 (d) 5 (e) 4 (f) 1 (g) 3
1.64 (a) 72 (b) 8.3 (c) 0.087 (d) 0.50 (e) 3.8
1.66 (a) 130.0 (b) 8.6 (c) 0.33
1.68 The answer is 497. The number with the fewest significant figures is 0.000450, which has three.
1.70 Gas: air, helium, oxygen Liquid: water, gasoline, olive oil Solid: ice, sugar, wood
1.72 56 kg, assuming that the room is empty
1.74 96°C
1.76 (c)
1.78 (a) mercury (b) 162 degrees (c) carbon
1.80 The quart is the better buy.

1.82 See Section 1.1.
1.84 (b)

*Chapter 2. Atoms*

2.4 Elements: (a)(g)(j) Compounds: (d)(e)(i)(k)
Mixtures: (b)(c)(f)(h)
2.6 A compound has a constant composition; a mixture does not.
2.8 See Section 2.3.
2.10 Only statement 4 (atoms combine to form molecules) is regarded as completely valid, and even here some atoms do not form molecules, but form ions instead.
2.12 (a) 1 (b) 1/1835 (c) 1
2.14 (a) 151 (b) 82 (c) 204 (d) 31 (e) 244
2.16 (a) 8 (b) 44 (c) 126 (d) 14
2.18 (a) 35 (b) 22 (c) 104 (d) 7 (e) 128 (f) 238
2.20 (a) oxygen (b) potassium (c) copper
(d) sodium (e) phosphorus (f) silver
(g) carbon (h) tin (i) iron
2.22 (a) an ion is a particle with an unequal number of protons and electrons (hence it has a charge).
(b) Isotopes are atoms with the same number of protons, but different numbers of neutrons.
2.24 (a) 16e, 16p, 16n (b) 18e, 17p, 20n
(c) 23e, 25p, 30n (d) 80e, 80p, 122n
(e) 10e, 13p, 14n
2.26 63.62 amu
2.28 Br, Cl, F; Rb, K, Li; He, Ne, Kr; Ca, Sr
2.30 (a) germanium (b) selenium (c) silicon
(d) aluminum
2.32 (a) spherical (b) dumbbell-shaped
2.34 See Section 2.6.
2.36 In all three cases the outer-shell configurations are the same, except for the shell number:
(a) Rb: $5s^1$ Li: $2s^1$ (b) O: $2s^22p^4$ Te: $4s^24p^4$
(c) Ge: $4s^24p^2$ Si: $3s^23p^2$
2.38 (a) F: $1s^22s^22p^5$; $F^-$: $1s^22s^22p^6$
(b) S: $1s^22s^22p^63s^23p^4$; $S^{2-}$: $1s^22s^22p^63s^23p^6$
(c) P: $1s^22s^22p^63s^23p^3$; $P^{3-}$: Same as $S^{2-}$
2.40 $7s^1$
2.42 because elements in the same column have the same electron configuration in their outer shells
2.44 lowest-to-highest: K, Li, Al, B, C, Cl, O, N, Ar
2.46 See Box 2A.
2.48 copper
2.50 irritation of the mucous membranes of the eye, nose, throat, and respiratory tract

2.52 (a) $s^2p^1$ (b) $s^2p^5$ (c) $s^2p^3$

2.54 (a) the state in which all electrons of an atom, ion, or molecule are in their lowest possible energy levels (b) all states in which one or more electrons are not in their lowest possible energy levels

2.56 (a) Ne (b) $K^+$ (c) $Br^-$ (d) $Ba^{2+}$ (e) $H^-$ (f) $Sc^{3+}$

2.58 $6.0 \times 10^{21}$ protons

2.60 Any state with an energy higher than the ground state. An atom can be in only one unique excited state at any time.

2.62 (a) N (b) Na (c) P (d) K (e) Sn (f) Zn (g) Ag (h) Br (i) C (j) Ca

2.64 $1s^22s^12p^1$, $1s^22s^13s^1$, etc.

2.66 (a) $1s$ (b) $2s2p$ (c) $3s3p3d$ (d) $4s4p4d4f$

## *Chapter 3. Chemical Bonds*

3.12 ionic and covalent

3.14 (a) gain 1 (b) lose 1 (c) lose 2 (d) gain 3 (e) none (f) gain 3 (g) gain 1 (h) gain or lose 4 (i) gain or lose 1 (j) gain 2

3.16 (a), (d), (f), and (k)

3.18 A charge of +6 is far too concentrated for a small ion like $O^{6+}$.

3.20 Yes, ions of transition elements (see Section 3.2).

3.22 See Section 3.3.

3.24

| | | | | |
|---|---|---|---|---|
| $NaCl$ | $Na_2SO_4$ | $Na_2CO_3$ | $NaHCO_3$ | $Na_3PO_4$ |
| $MgCl_2$ | $MgSO_4$ | $MgCO_3$ | $Mg(HCO_3)_2$ | $Mg_3(PO_4)_2$ |
| $AlCl_3$ | $Al_2(SO_4)_3$ | $Al_2(CO_3)_3$ | $Al(HCO_3)_3$ | $AlPO_4$ |
| $NH_4Cl$ | $(NH_4)_2SO_4$ | $(NH_4)_2CO_3$ | $NH_4HCO_3$ | $(NH_4)_3PO_4$ |
| $FeCl_3$ | $Fe_2(SO_4)_3$ | $Fe_2(CO_3)_3$ | $Fe(HCO_3)_3$ | $FePO_4$ |

| | |
|---|---|
| $Na_2HPO_4$ | $NaH_2PO_4$ |
| $MgHPO_4$ | $Mg(H_2PO_4)_2$ |
| $Al_2(HPO_4)_3$ | $Al(H_2PO_4)_3$ |
| $(NH_4)_2HPO_4$ | $NH_4H_2PO_4$ |
| $Fe_2(HPO_4)_3$ | $Fe(H_2PO_4)_3$ |

3.26 (a) $Na^+$ $Br^-$ (b) $Fe^{2+}$ $SO_3^{2-}$ (c) $Mg^{2+}$ $PO_4^{3-}$ (d) $K^+$ $H_2PO_4^-$ (e) $Na^+$ $HCO_3^-$ (f) $Ba^{2+}$ $NO_3^-$

3.28 See Section 3.4.

3.30 See Section 3.4.

3.32 (a) The left C has five bonds. (b) correct (c) The second C has three bonds; the O has one bond. (d) The F has two bonds. (e) correct (f) The second C has five bonds.

3.34 Oxygen only forms two bonds, so it cannot form a triple bond.

3.36 A molecular formula gives ratios of atoms; a structural formula shows which atoms are connected to which; a Lewis structure shows all outer-shell electrons.

3.38 (a) 8 (b) 18 (c) 24 (d) 24 (e) 32

3.40 (a) H—C≡N: (b) H—O—N=O: (c)

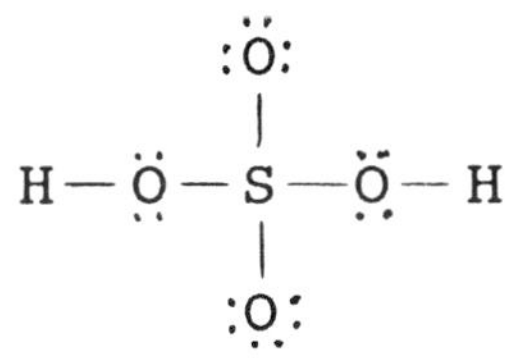

(d)

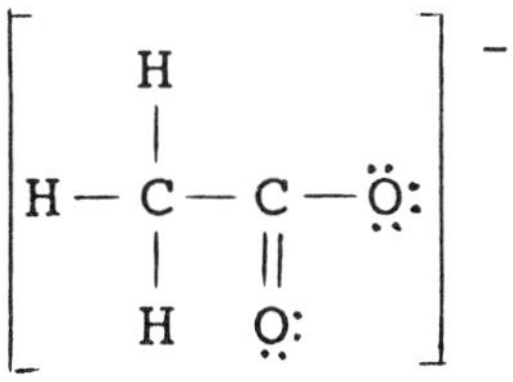

(e)

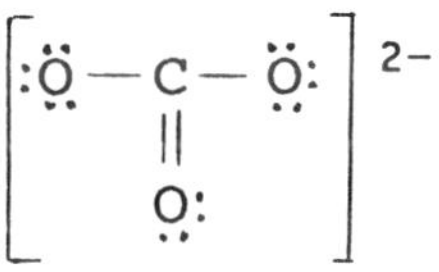

(f)

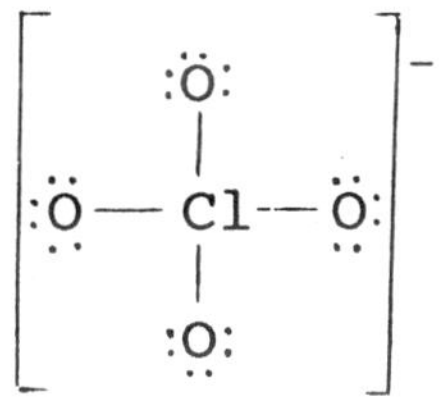

3.42 :C≡O: ; yes

3.44 These are completely different species whose Lewis structures are: chlorine atom :Cl· chlorine molecule :Cl—Cl: chloride ion :Cl: $^-$

3.46 (a) linear (b) trigonal (c) tetrahedral

3.48 (a) Cl (b) N (c) O (d) Cl (e) O (f) S (g) O

3.50 Most polar: (b) Least polar: (c)

3.52 Yes. The negative charge is at the S atom. The positive charge is halfway between the hydrogens.

3.54 Ionic: (b)(e)(f); Polar covalent: (a)(d)(h)(i); Nonpolar: (c)(g)

3.56 (a)

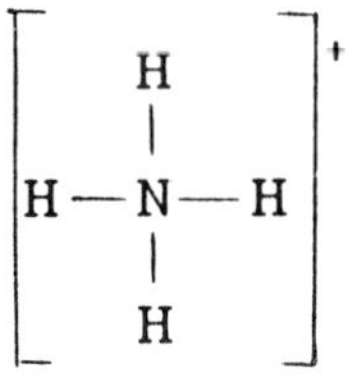

Cl $^-$

(b) [H—O:]$^-$ Na$^+$

(c)

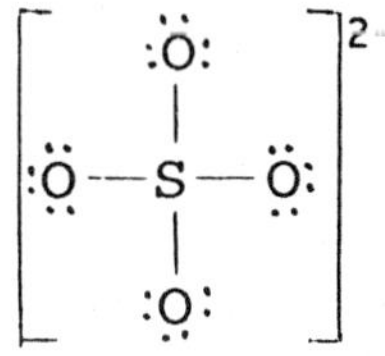

2K$^+$

(d)

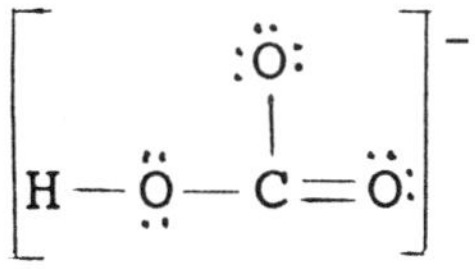

K$^+$

(e)

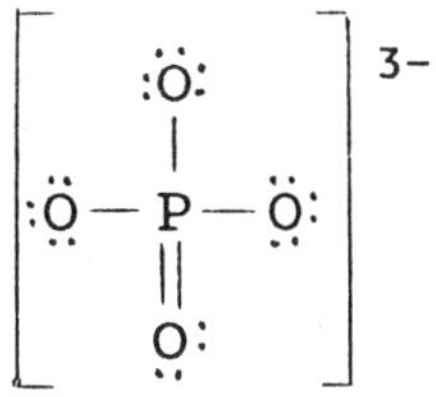

3Na$^+$

3.58 (a) KBr (b) CaO (c) $Mg(CH_3COO)_2$ (d) $Li_2SO_4$
(e) $NH_4HSO_3$ (f) $Cu_3(PO_4)_2$ (g) $Ag_2CO_3$
(h) $Pb(OH)_2$ (i) $Ba(MnO_4)_2$

3.60 (a) CO (b) BrCl (c) $PCl_5$ (d) $SiBr_4$
(e) $N_2O_3$ (f) $SF_6$

3.62 bone formation

3.64 $K^+$, $HPO_4^{2-}$

3.66 $Fe^{2+}$

3.68 an external anti-infective

3.70 tetrahedral

3.72 (a) sodium sulfite (b) potassium permanganate (c) strontium cyanide (d) iron(II) oxide or ferrous oxide (e) ammonium carbonate (f) lithium nitrite (g) sodium bicarbonate

3.74 Compounds (b) (c) (e) and (f) are polar.

3.76 (a) $K_2S$ (b) $Al(NO_3)_3$ (c) $PbSO_3$ (d) $Ca(HCO_3)_2$
(e) $Hg(H_2PO_4)_2$ (f) $BaH_2$

3.78

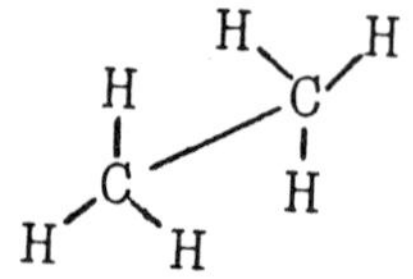

3.80 (a) nonmetals (b) metals

*Chapter 4. Chemical Reactions*

4.14 (a) 159.8 amu (b) 20.2 amu (c) 253.8 amu
(d) 123.9 amu (e) 256.5 amu

4.16 (a) 3.06 moles $CH_4$ (b) 10.8 moles acetone
(c) 0.17 mole $KH_2PO_4$ (d) 0.0626 mole quinine

4.18 (a) 32.5 moles H atoms (b) 1.48 moles C atoms
(c) $7.3 \times 10^5$ moles O atoms
(d) 1.99 moles Ni atoms (e) $1.3 \times 10^{-5}$ moles N atoms

4.20 The same; that is, just about 2:1

4.22 $5.7 \times 10^{-22}$ g

4.24 $6.1 \times 10^{18}$ molecules of cholesterol

4.26 (a) $Cl_2 + H_2 \longrightarrow 2HCl$
(b) $2H_2 + O_2 \longrightarrow 2H_2O$
(c) $2K + Br_2 \longrightarrow 2KBr$
(d) $MgCO_3 \longrightarrow CO_2 + MgO$

4.28 (a) the study of weight relationships in chemical reactions (b) a solution in which the solvent is water

4.30 7.34 moles $H_2$

4.32 328 g $C_2H_4O$

4.34 7.5 g $CO_2$

4.36 0.13 g C

4.38 150 g of trichloroaniline

4.40 85.6%

4.40 85.6%

4.42 (a) $K^+$ and $Cl^-$ (b) $Ba^{2+}$ + $SO4^{2-}$ $\longrightarrow$ $BaSO_4(s)$

4.44 $Pb^{2+}$ + $S^{2-}$ $\longrightarrow$ $PbS_2(s)$

4.46 $NH_4^+$ + $OH^-$ $\longrightarrow$ $NH_3(g)$ + $H_2O$

4.48 Soluble: (a)(b)(f)(g)(h)(l)
Insoluble: (c)(d)(e)(i)(j)(k)

4.50 Oxidation is (1) loss of electrons (2) gain of oxygen or loss of hydrogen. Reduction is (1) gain of electrons (2) loss of oxygen or gain of hydrogen.

4.52 (a) Sn gets oxidized; $Au^+$ gets reduced.
(b) $Au^+$ is the oxidizing agent; Sn the reducing agent.

4.54 (a) An exothermic process is one that gives off heat.
(b) An endothermic process is one that requires heat.

4.56 This reaction is endothermic and absorbs 23 800 cal.

$$2SO_3 \longrightarrow 2SO_2 + O_2 - 23\ 800\ \text{cal}$$

4.58 88.7 g iron

4.60

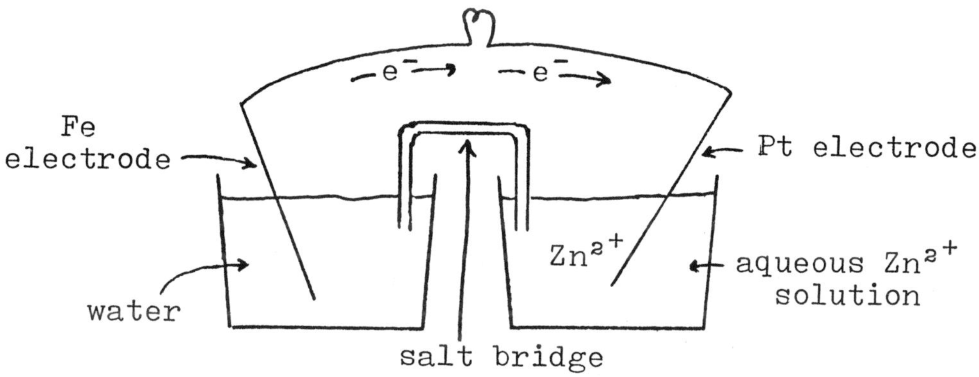


4.62 See Box 4D.

4.64 (a) 0.64 mole (b) 0.96 mole (c) 1.2 moles

4.66 an ion that does not participate in a reaction

4.68 (a) 19.2 moles $Fe_2O_3$ (b) 1610 g CO

4.70 (a) An exergonic process is one that gives off energy in any form, not necessarily heat.
(b) An endergonic process is the opposite.

4.72 $2As^{3+}$ + $3S^{2-}$ $\longrightarrow$ $As_2S_3(s)$

4.74 $H_2$ is in excess.

*Chapter 5. Gases, Liquids and Solids*

5.8 They are directly related. Doubling the Kelvin temperature

5.10

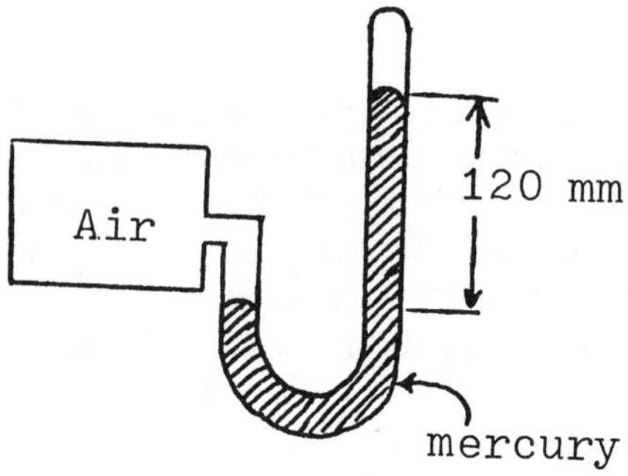


5.12 (a)

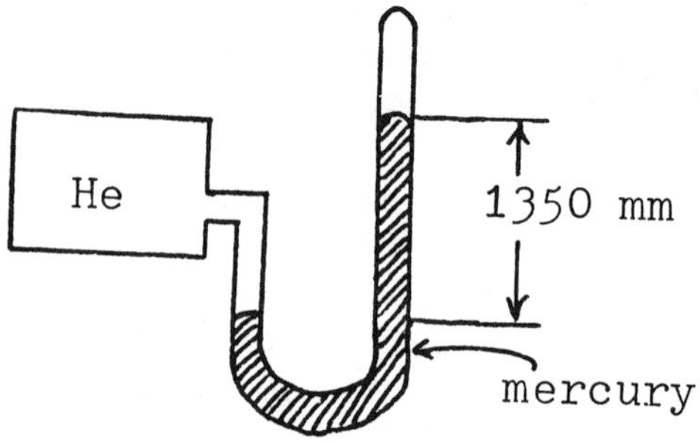


(b) The level on the left side falls and the level on the right side rises.
(c) The level on the left side rises and the level on the right side falls.

5.14 stays the same
5.16 342°C
5.18 0.822 atm
5.20 6.1 atm
5.22 4.35 atm
5.24 4930 L
5.26 1.97 atm
5.28 (a) 2.33 moles (b) no
5.30 2.8 L
5.32 438 K
5.34 (a) 24 moles (b) 770 g
5.36 $SO_2$: 2.86 g/L $CH_4$: 0.714 g/L $F_2$: 1.70 g/L
5.38 $2.68 \times 10^{24}$ molecules
5.40 Extensive: <u>n</u>, <u>V</u> Intensive: <u>P</u>, <u>T</u>
5.42 324 mm Hg
5.44 (b)
5.46 See Sec. 5.9.
5.48 See Sec. 5.11.
5.50 (a)
5.52 (a) London dispersion forces (b) dipole-dipole interactions (c) hydrogen bonds
5.54 (a) 75 mm Hg (b) 15 mm Hg (c) 600 mm Hg
5.56 (a) $HCl < HBr < HI$ Boiling points increase with molecular weight (b) $O_2 < HCl < H_2O_2$ Here the molecular weights are about the same, but the strength of interaction increases: London $<$ dipole $<$ hydrogen bond.

5.58 When the temperature decreases the kinetic energy decreases; it is not lost but converted to potential energy or energy of interaction. More interaction means more attraction and hence condensation.

5.60 conversion of a solid directly to a gas

5.62 Because mercury vapors are poisonous. If spilled and not cleaned up, some of the mercury will go into the vapor phase, and people will inhale it.

5.64 The volume of the chest cavity increases, lowering the pressure. Air flows in from outside.

5.66 highest in venous blood, lowest in arterial blood

5.68 It is the hardest known substance.

5.70 $NH_3$

5.72 CO, because it has a higher dipole moment.

5.74 (c) water

5.76 See Sec. 5.11.

5.78 There is a lot of empty space between gas molecules, so electromagnetic radiation (including visible light) passes through without much interaction.

5.80 In evaporation a liqud is converted to a gas; in sublimation a solid is converted directly to a gas without going through the liquid state.

## *Chapter 6. Solutions and Colloids*

6.12 A true solution is clear and transparent; a colloidal dispersion is cloudy.

6.14 (a) no (b) taste it

6.16 (a) wine, vinegar (b) honey, tea
(c) oxygen in rivers, household ammonia
(d) air, laughing gas-air mixture (anesthetic)

6.18 If they are completely soluble in each other, they cannot form a supersaturated solution.

6.20 Gasoline: (a)(e) Water: (b)(c)(d)

6.22 (1) Heat the mixture. (2) Add more water.

6.24 By adding a grain of crystal to a supersaturated solution a solid surface is provided upon which solute molecules can converge.

6.26 The beer is bottled under pressure. The solubility of $CO_2$ in water increases with increasing pressure. When the bottle is opened, the pressure decreases to 1 atm, decreasing the solubility of $CO_2$. Hence the beer loses $CO_2$ (goes flat).

6.28 (a) Dissolve 86.5 g KCl in 163.5 g water.
(b) Dissolve 6.24 g NaOH in 149.76 g water.
(c) Dissolve 0.96 g formic acid in 299.04 g water.

6.30 In each case, to the given weight of solute add enough water to make the total volume specified.
(a) 9.0 g NaCl; 250 mL total volume
(b) 31 g glycine; 625 mL total volume

(c) 5.96 g $Na_2SO_4$; 43.5 mL total volume
(d) 10.9 g acetone; 518 mL total volume

6.32 (a) 1.3 M (b) 0.306 M (c) 0.023 M
6.34 68 mL
6.36 0.121 M
6.38 0.109 M
6.40 (a) 78.8 g (b) 48.4 g (c) 0.20 g
6.42 To 530 mL of 95% ethyl alcohol add enough fruit juice to make 5.0 L of fruit punch.
6.44 200 ppb
6.46 The water molecules surround the cation with their negative ends turned toward the positive ion.
6.48 Strong: (a)(c) Weak: (e) Non: (b)(d)
6.50 The water molecule forms hydrogen bonds with the OH group of ethanol.
6.52 (a) solid in liquid (b) liquid in solid (c) liquid in liquid (d) gas in liquid
6.54 (a) turbidity, because it is a colloid
(b) the Tyndall effect
6.56 Brownian motion is the random motion of colloidal particles. The higher the temperature, the faster the motion. the faster the motion, the more collisions, and more aggregation of the particles, which will eventually settle out.
6.58 -7.35°C
6.60 79 g salt
6.62 See Sec. 6.9.
6.64 (a) compartment B (b) compartment B
6.66 See Sec. 6.9.
6.68 carbon dioxide in the air
6.70 sulfur dioxide
6.72 See Box 6B.
6.74 Hypoglycemia shifts $K^+$ from the extracellular fluid into the cells; $K^+$ controls nerve transmission to the heart muscle. With lowered $K^+$ concentration this control is impaired.
6.76 Anything greater than 3.8% $MgSO_4 \cdot 7H_2O$, which is isotonic with 0.9% NaCl.
6.78 The alcohol removes the solvation layer that protects the colloid.
6.80 0.31 osmol
6.82 $2.44 \times 10^{-5}$ ppm, 0.0244 ppb
6.84 0.006 g
6.86 In cold water the starch particles are larger than 1000 nm. In boiled water the particles are smaller: between 1 and 1000 nm.
6.88 (a)
6.90 The solubility of most substances increases with increasing temperature.

7.4 The change in concentration of a reactant or product per unit time. Moles per liter per minute.

7.6 $2.40 \times 10^{-3}$ mole/L·min

7.8 1. They don't have the activation energy.
2. They don't have the proper orientation.

7.10 The breaking of chemical bonds is always endothermic, because bonded atoms always have a lower energy than nonbonded atoms.

7.12

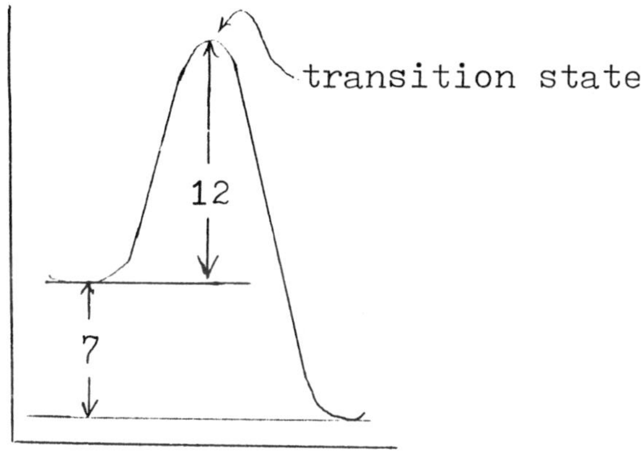


7.14 The higher the concentration, the more molecules there are in the same volume, so there are more collisions per unit time.

7.16 As the reaction goes on, the concentrations of reactants decrease. In most cases the rate is directly proportional to the concentrations of the reactants.

7.18 (1) The molecules move faster, causing more collisions per unit time. (2) There is a higher percentage of effective collisions.

7.20 In a solid, only the surface molecules (or ions) can react. Small particles have a larger surface area per unit mass than large particles. Therefore the larger the surface area per gram, the faster the reaction.

7.22 No; no. We can start with any proportions of reactants, so at equilibrium they won't have equal concentrations.

7.24 (a) $COCl_2 \rightleftharpoons CO + Cl_2$

(b) $CO + Cl_2 \rightleftharpoons COCl_2$

(c) $2HBr \rightleftharpoons H_2 + Br_2$

(d) $3F_2 + PH_3 \rightleftharpoons 3HF + PF_3$

7.26 $\underline{K} = 172$

7.28 No. Equilibrium would be reached very quickly, but at equilibrium there would be virtually no $NH_2^-$ or $H_3O^+$.

7.30 the set of conditions at which $\underline{K} = 1.72$

7.32 (a) left (b) left (c) right (d) right (e) no shift

7.34 (a) left (b) right (d) left (e) right (e) no shift

7.36 Yes. A change in temperature

7.38 eliminate static, which provides sparks; use antistatic ceramic tiles or PVC

7.40 Fever enables the body to kill germs faster, by mobilizing the immune defense mechanism.

7.42 See Box 7C.

7.44 $\underline{K} = 50.$

7.46 At -5° the rate is 0.70 moles of HCl per liter per second; at 45°C the rate is 22.4 moles of HCl per liter per second.

7.48 0.026 $\underline{M}$

7.50 (a) $\underline{K} = \dfrac{[OH^-]^2[H_2PO_4^-]}{[PO_4^{3-}][H_2O]^2}$ (b) $\underline{K} = \dfrac{[H_2O][HNO_3]}{[H_3O^+][NO^{3-}]}$

(c) $\underline{K} = \dfrac{[H_2O]^2}{[OH^-][H_3O^+]}$

7.52 $2.0 \times 10^{-3}$ mole of $N_2O_4$ per liter per second

7.54 The activation energy must be less.

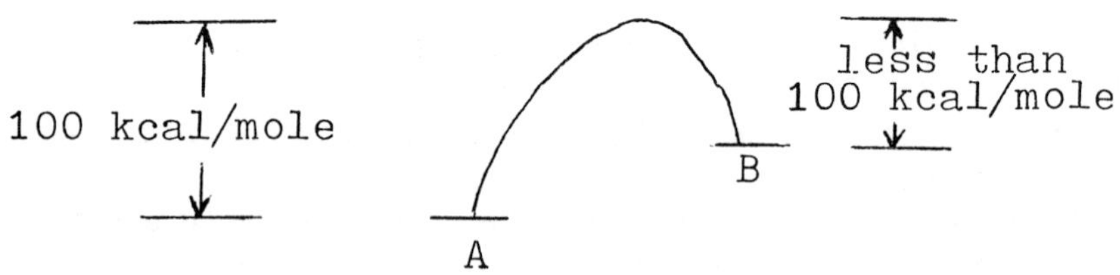


7.56

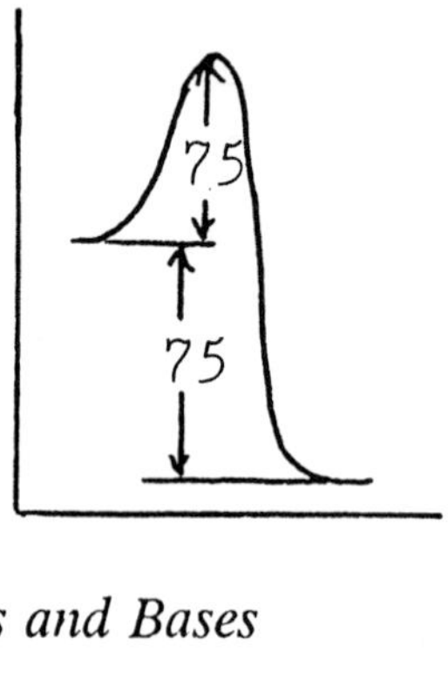


*Chapter 8. Acids and Bases*

8.16 $NH_3 + H_2O \rightleftharpoons OH^- + NH_4^+$

$CH_3NH_2 + H_2O \rightleftharpoons OH^- + CH_3NH_3^+$

8.18 (a) $HSO_4^-$ (b) $Br^-$ (c) $H_2PO_4^-$ (d) $CO_3^{2-}$
(e) $NH_3$ (f) $OH^-$ (g) $H_2O$ (h) $HPO_4^{2-}$
8.20 Monoprotic: (a)(c) Diprotic: (b)(e) Triprotic: (d)
8.22 (a)(c)(d)
8.24 They react with (neutralize) each other, forming water and salts.
8.26 (a) $10^{-3}$ M (b) $10^{-11}$ M (c) $10^{-7}$ M (d) $10^{-15}$ M
8.28 (a) 4 (b) 9 (c) 3 (d) 7 (e) 0
acidic: (a)(c)(e) basic: (b) neutral: (d)
8.30 (a) 9.7 (b) 2.3 (c) 4.1 (d) 6.5
acidic: (b)(c)(d) basic (a)
8.32 (a) $10^{-2}$ (b) $10^{-13}$ (c) $10^{-10}$ (d) $10^{0}$ (e) $10^{-7}$
acidic: (b)(c) basic: (a)(d) neutral: (e)
8.34 $NO_3^-$ does not react with water, but $Ca^{2+}$ does:

$$Ca^{2+} + 4H_2O \rightleftharpoons Ca(OH)_2 + 2H_3O^+$$

The $H_3O^+$ ions make the solution acidic.

8.36 Acidic: (d)(e)(g) Basic: (a)(b) Neutral: (c)(f)(h)

8.38 (a) $H_3O^+ + CH_3COO^- \rightleftharpoons CH_3COOH + H_2O$

(b) $OH^- + CH_3COOH \rightleftharpoons CH_3COO^- + H_2O$

8.40 3.45 ($pK_a$ of hydrofluoric acid)
8.42 (a) 4.85 (b) The 1 M solution has ten times the buffer capacity of the 0.1 M solution.
8.44 to determine the acid or base concentration of a solution
8.46 (a) 1.29 (b) 0.822 (c) 3.80 (d) 0.0942
8.48 0.088 N
8.50 (a) To 12 g of solid NaOH add enough water to make a total volume of 400.0 mL.
(b) To 6.1 g of solid $Ba(OH)_2$ add enough water to make a total volume of 1.0 L.
8.52 0.57 N
8.54 0.900 N
8.56 (a) 3.2 M (b) 9.0 M (c) 0.066 M (d) 0.73 M
(e) 0.035 M
8.58 magnesium hydroxide
8.60 strong bases
8.62 Alka-Seltzer, Bromo-seltzer, Chooz, Riopan, Tums
8.64 sodium bicarbonate
8.66 See Box 8E.
8.68 Solving the Henderson-Hasselbach equation:

$$10.6 = 9.9 + \log(\text{phenoxide}/\text{phenol})$$

we obtain a phenoxide: phenol ratio of 5:1.
We therefore make up a solution that contains 5 times as many moles of phenoxide ion as of phenol.

8.70 0.71 M

8.72 $NH_3 + HCl \longrightarrow NH_4^+ + Cl^-$

All the $NH_3$ will be converted to $NH_4^+$ by the above reaction. None will remain.

8.74 0.900 M

8.76 11.2

8.78 (a) $H_3O^+ + F^- \rightleftharpoons HF + H_2O$
(b) $OH^- + HF \rightleftharpoons F^- + H_2O$

8.80 (a) 0 (b) 1 x $10^0$ = 1 mole/liter

8.82 7.9 x $10^{-2}$ M

*Chapter 9. Nuclear Chemistry*

9.4 By accident, a photographic paper placed near a radioactive rock became blackened.

9.6 5.0 x $10^{-5}$ cm

9.8 See Section 9.1.

9.10 Longest wavelength: (a) Highest energy: (c)

9.12 (a) ${}^{1}_{1}H$ (b) ${}^{16}_{8}O$ (c) ${}^{40}_{20}Ca$ (d) ${}^{28}_{14}Si$

9.14 Europium-151

9.16 ${}^{51}_{23}V$

9.18 a neutron ${}^{27}_{13}Al + {}^{4}_{2}He \longrightarrow {}^{30}_{15}P + {}^{1}_{0}n$

9.20 thorium-230

9.22 gamma rays

9.24 a diagram that shows how much radioactive material is left after a certain time

9.26 63 mg

9.28 There are no ways.

9.30 (a) There will be little change after 2 h. The activity will be 24 mCi and 8.9 x $10^8$ counts per minute.
(b) 3 mCi will remain in the thyroid, assuming none has been excreted.

9.32 because the alpha particle knocks out an electron from atoms or molecules, turning them into ions

9.34 0.63 $cm^3$

9.36 11 percent

9.38 alpha particles

9.40 (a) cobalt-60 (b) selenium-75 (c) strontium-85
(d) carbon-11, technetium-99m (e) mercury-197, technetium-99m

9.42 the 14.3 kcal/mole gamma radiation

9.44 Iodine-131 concentrates in the thyroid. Yttrium-90 concentrates in the pituitary gland.

9.46 helium

9.48 a neutron (${}^{1}_{0}n$)

9.50 (a) 4 (b) 1 (c) 2 (d) 3

9.52 2 spots (the neutrons and gamma rays have no charge)

9.54 The three rays have different charges: positive, negative, and neutral.
9.56 about 5730 years
9.58 about 6 percent
9.60 soft tissues such as brain
9.62 See Box 9G.
9.64 500 rad
9.66 (b)
9.68 No, we count the ionized gas molecules produced by the beta particles.
9.70 x-rays
9.72 A neutron is converted to a proton by emitting an electron.
9.74 28 s
9.76 It is eliminated from the body long before the 602 days have passed.
9.78 the lowest possible energy state

*Chapter 10. Organic Chemistry. Alkanes*

10.8 (a)(b)(e)
10.10 (a) $CH_3(CH_2)_4CH_3$ (b) KBr (c) NaOH
10.12 (a) a compound whose origin and properties have been published in a scientific journal
(b) by extraction from a natural source and by synthesis
10.14 They have different structures, and properties depend on structure, not on molecular formula.
10.16 (a)

```
CH3-CH-Br          Br-CH2CH2-Br
    |
    Br
```

(b)

```
CH3CH2CH2CH2-Cl           CH3CH2-CH-CH3
                                 |
                                 Cl
                             CH3
                             |
CH3-CH-CH2-Cl           CH3-C-Cl
    |                        |
    CH3                      CH3
```

10.18 (a) The isomers (all $C_5H_{10}O$) are (a)(b)(c)(d)(f)(g)(h)
(b)

```
CH3-C-CH2CH2CH3,   CH3-CH=CH-CH2CH2-OH, and
    ||
    O

CH3CH2-C-CH2CH3 are three of them. There are many others.
       ||
       O
```

10.20 (a) O has 1 bond, one of the H's has 2 bonds
(b) C has 3 bonds (c) all correct
(d) the first and fourth C have 5 bonds
(e) O has 3 bonds, N has 1 bond
(f) all correct

10.22 (a)

10.24 (a) Each of the 4 $sp^3$ orbitals of the carbon overlaps with a $p$ orbital of one bromine.
(b) One $sp^3$ orbital of the carbon overlaps with an $s$ orbital of the hydrogen. Each of the other 3 $sp^3$ orbitals of the carbon overlaps with a $p$ orbital of a fluorine.
(c) Each carbon uses 4 $sp^3$ orbitals. One $sp^3$ orbital of one carbon overlaps with one $sp^3$ orbital of the other. One $sp^3$ orbital of one carbon overlaps with the $p$ orbital of the bromine. Each of the other 5 $sp^3$ orbitals overlaps with the $s$ orbital of a different hydrogen.

10.26 The real molecule has a zig-zag geometry. All the angles are about 109.5°.

10.28 (a) none (b) The ring carbon attached to both the $CH_3$ and the $CH_2CH_3$ groups is quaternary. The ring carbon attached only to the $CH_3$ group is tertiary. (c) Counting from right to left along the straight chain, carbon 2 is quaternary; carbons 3 and 4 are tertiary.

10.30 (a)

(1) $CH_3-CH_2-CH_2-Cl$ 1-chloropropane

(2) $CH_3-CH-CH_3$ (Cl on the central CH) 2-chloropropane
$\quad\quad |$
$\quad\quad Cl$

(b)

(1) $CH_3-CH_2-CH_2-CH_2-CH_2-Br$ 1-bromopentane

(2) $CH_3-CH_2-CH_2-CH-CH_3$ 2-bromopentane
$\quad\quad\quad\quad\quad |$
$\quad\quad\quad\quad\quad Br$

(3) $CH_3-CH_2-CH-CH_2-CH_3$ 3-bromopentane
$\quad\quad\quad\quad |$
$\quad\quad\quad\quad Br$

(4) $CH_3-CH_2-CH-CH_2-Br$ 1-bromo-2-methylbutane
$\quad\quad\quad\quad |$
$\quad\quad\quad\quad CH_3$

(5) 2-bromo-2-methylbutane
$\quad\quad\quad\quad CH_3$
$\quad\quad\quad\quad |$
$CH_3-CH_2-C-CH_3$
$\quad\quad\quad\quad |$
$\quad\quad\quad\quad Br$

(6) $CH_3-CH-CH-CH_3$ 2-bromo-3-methylbutane
$\quad\quad |\quad |$
$\quad\quad Br \; CH_3$

10.30 continued

(7) 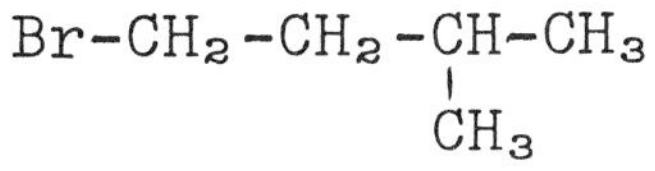 1-bromo-3-methylbutane

(8) 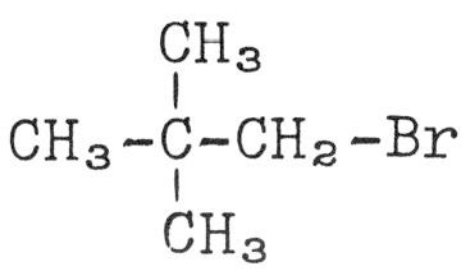 1-bromo-2,2-dimethylpropane

10.32 (a) 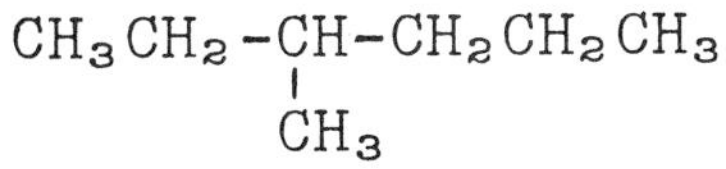

(b) 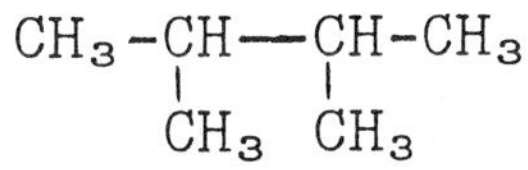

(c) 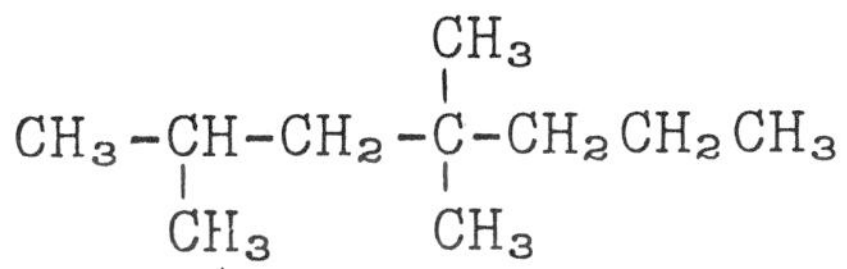

(d) 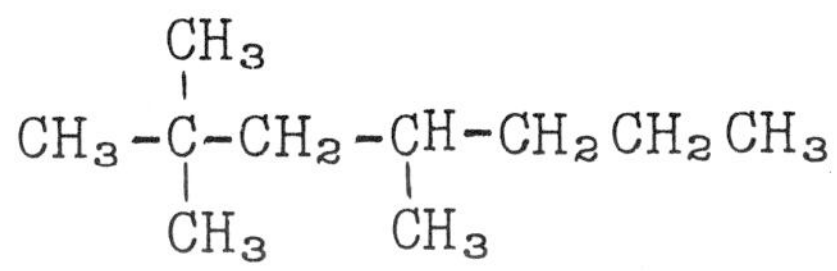

(e) 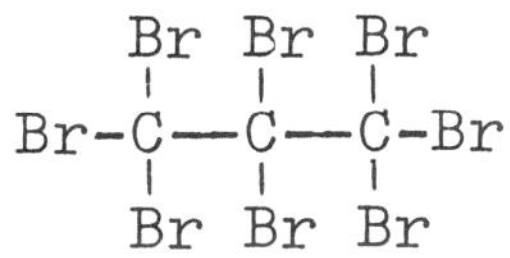

(f) 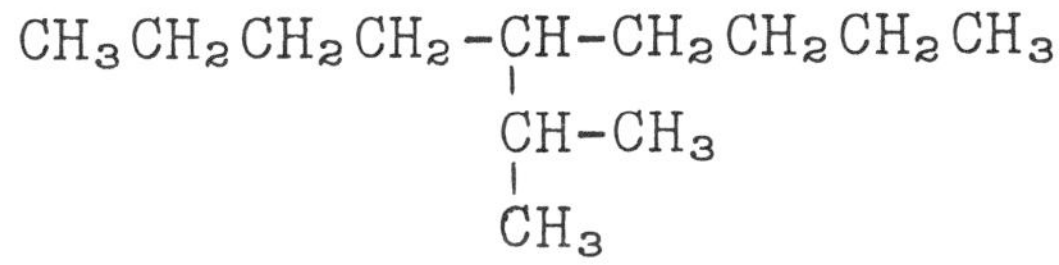

(g) 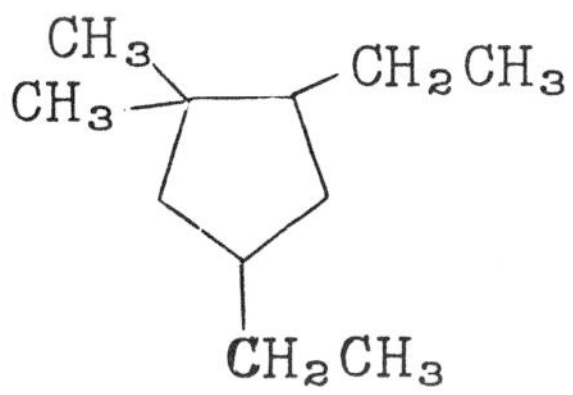

10.34 (a) ethyl (b) *tert*-butyl (c) isobutyl (d) chloro
(e) isopropyl (f) pentyl or *n*-pentyl

10.36 (a) cyclobutane (b) 1,3-dimethylcyclopentane
(c) 1,1,2,4-tetramethylcyclohexane
(d) 2-ethyl-1,3-dimethylcyclohexane
(e) *sec*-butylcyclopentane

10.38 (a) The prefix di requires two numbers. Only one is given.
(b) Pentane has five carbons. There cannot be a group only in the 4 position, since a lower number would be obtained counting from the other end.
(c) There cannot be a methyl group at the end of a parent chain, since that only makes the parent chain longer.
(d) There cannot be an ethyl group in the 2 position of an alkane, since that only makes the parent chain longer.
(e) If two numbers are given, the prefix di must be used.
(f) If there is only one group on a ring, it gets no number (the number 1 is understood).
(g) There cannot be a two-membered ring.
(h) Same answer as (d).

10.40 (1) 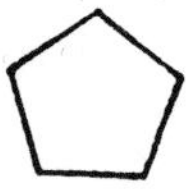 cyclopentane

(2) 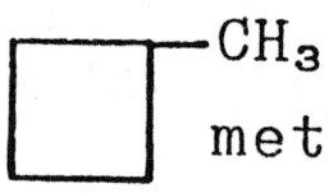 methylcyclobutane

(3) 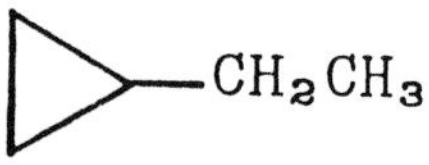 ethylcyclopropane

(4) 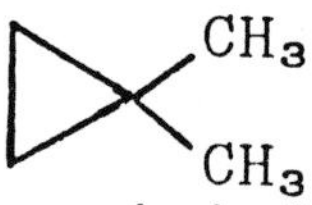 1,1-dimethylcyclopropane

(5) 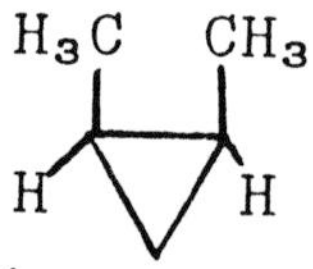 cis-1,2-dimethylcyclopropane

(6) 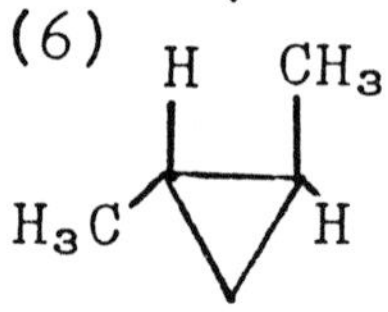 trans-1,2-dimethylcyclopropane

10.42 (a) 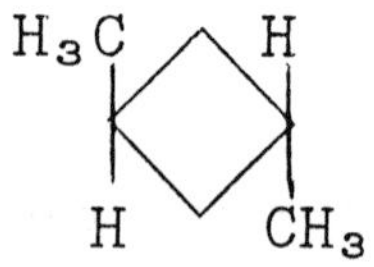 (b) 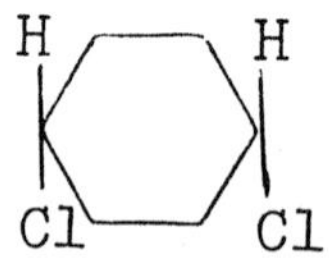

10.44 (a) liquid (b) no (c) yes (d) less dense

10.46 (a) $C_4H_8 + 6O_2 \longrightarrow 4CO_2 + 4H_2O$

(b) $C_7H_{16} + 11O_2 \longrightarrow 7CO_2 + 8H_2O$

(c) $2C_{10}H_{22} + 31O_2 \longrightarrow 20CO_2 + 22H_2O$

10.48 (a) $CH_2{=}CH_2$ (b) $CH_3{-}C({=}O){-}OH$ (c) $CH_3{-}NH_2$

(d) $CH_3{-}C({=}O){-}NH_2$ (e) $CH_3{-}OH$ (f) $CH_3{-}O{-}CH_3$

(g) $CH_3{-}SH$ (h) $CH_3{-}C({=}O){-}H$ (i) $CH_3{-}C({=}O){-}O{-}CH_3$

(j) $CH_3{-}C({=}O){-}CH_3$

10.50 (a) halide, alcohol (b) ketone, carboxylic ester (c) halide, double bond (d) ether, amine (e) alcohol (f) ketone, aldehyde

10.52 Cyclopropane has a low solubility in the blood, so it is eliminated rapidly.

10.54 Each molecule rapidly and spontaneously converts to the other.

10.56 See Box 10D.

10.58 CO binds to hemoglobin much more strongly than does $O_2$, so administration of $O_2$ will not dislodge it, and fresh blood must be added.

10.60 (a) 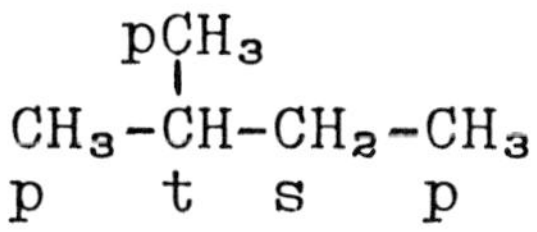 (b) 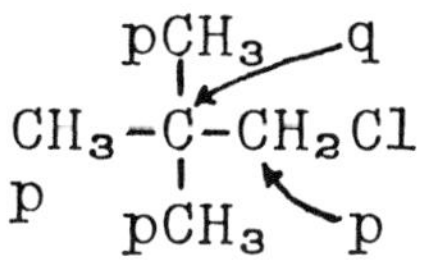

(c) 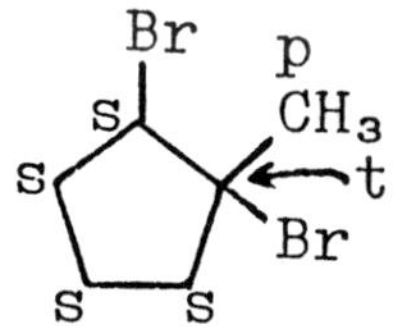 (d) $CH_3-CH_2-CH_3$
p s p

10.62 (1) 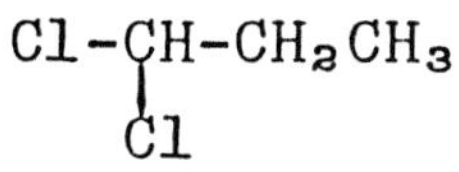 1,1-dichloropropane

(2) 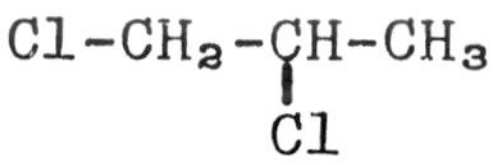 1,2-dichloropropane

(3) $Cl-CH_2CH_2CH_2-Cl$ 1,3-dichloropropane

(4) 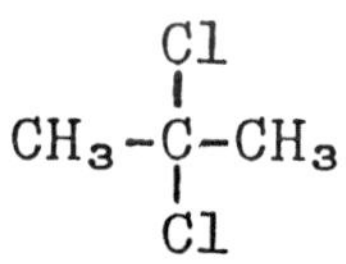 2,2-dichloropropane

10.64 (a) alcohol, ketone (b) carboxylic acid, carboxylic ester
(c) alcohol, aldehyde

10.66 (1) $CH_3(CH_2)_5CH_3$ heptane

(2) 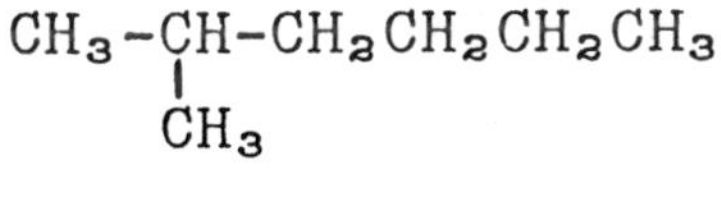 2-methylhexane

(3) 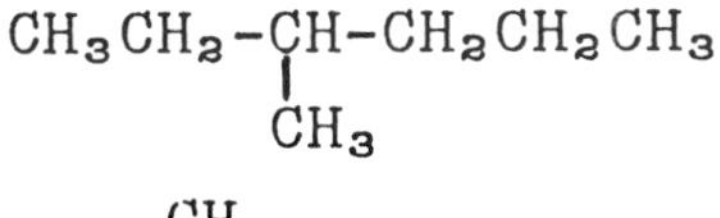 3-methylhexane

(4) 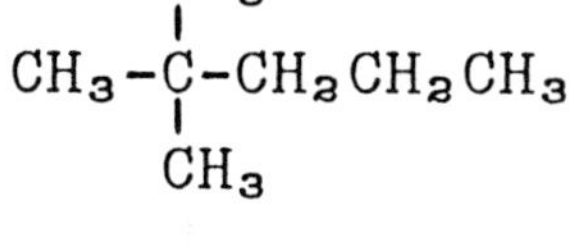 2,2-dimethylpentane

(5) 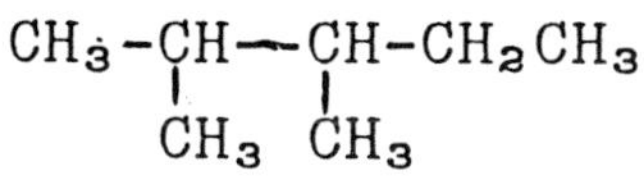 2,3-dimethylpentane

(6) 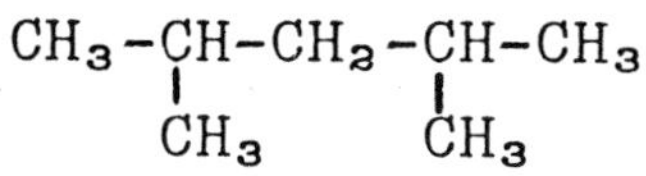 2,4-dimethylpentane

10.66 continued

(7) $CH_3CH_2-C(CH_3)_2-CH_2CH_3$ 3,3-dimethylpentane

(8) $CH_3CH_2-CH(CH_2CH_3)-CH_2CH_3$ 3-ethylpentane

(9) $CH_3-CH(CH_3)-C(CH_3)_2-CH_3$ 2,2,3-trimethylbutane

10.68 There is a much greater percentage increase in molecular weight in going from $C_1$ to $C_2$ or from $C_2$ to $C_3$ than from $C_8$ to $C_9$ or from $C_9$ to $C_{10}$.

*Chapter 11. Alkenes, Alkynes, and Aromatic Compounds*

11.14 Alkenes: (a)(e) Alkyne: (c)

11.16 (a) $H_2C=CH-CH_3$ (b) $H_2C=CH-CH_2CH_2CH_3$

(c) $CH_3-C(CH_3)=CH-CH(CH_3)-CH_3$ (d) $Br-CH_2CH_2-C(Br)=CH-CH(CH_3)-CH_2CH_2CH_3$

(e)

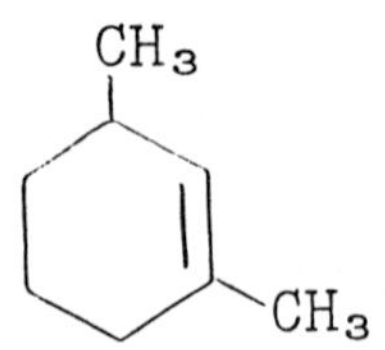


(f) $CH_3-CH=CH-CH=CH-CH_2CH_3$

(g)

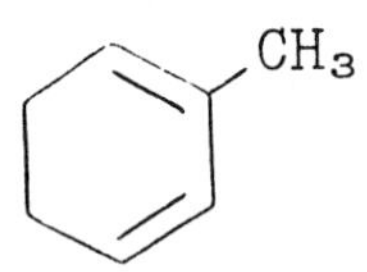


11.18 three

11.20 There are no cis and trans isomers in (a), (c), and (f).

(b)

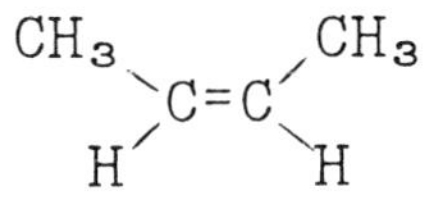

cis

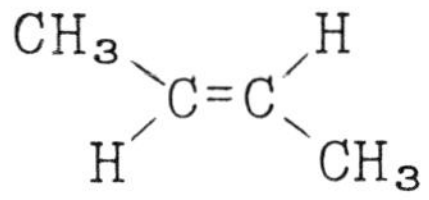

trans

(d)

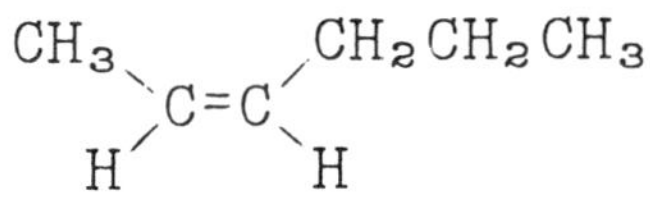

cis

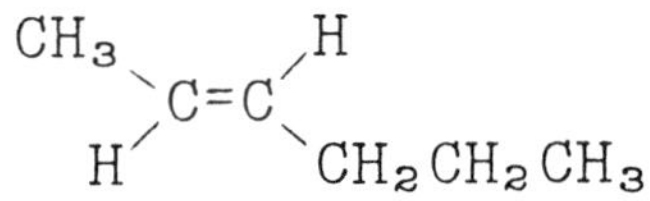

trans

(e)

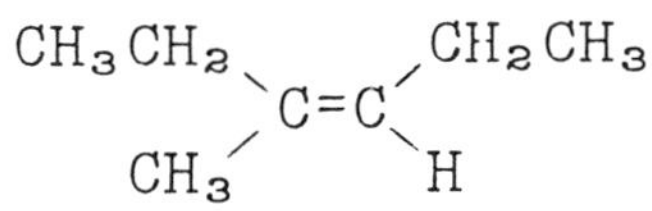

cis

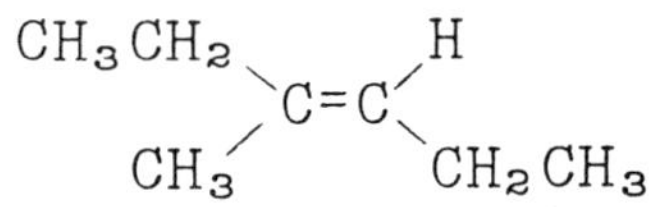

trans

(g)

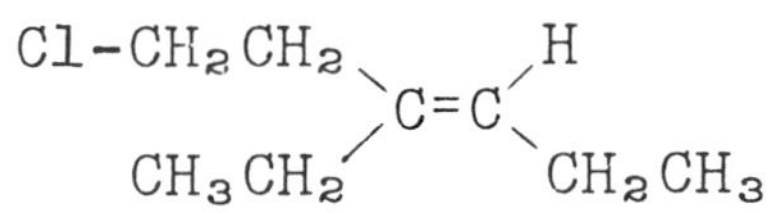

cis

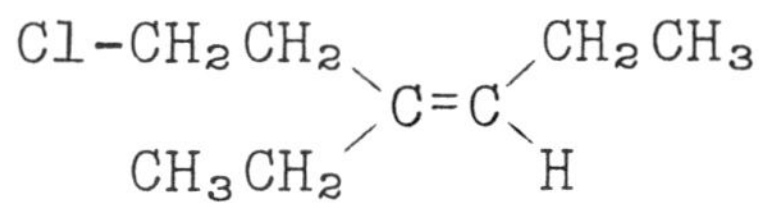

trans

11.22 (a) <u>cis</u>-2-pentene
(b) <u>trans</u>-1,4-dichloro-2-methyl-2-butene
(c) <u>cis</u>-2-chloro-3-heptene
(d) <u>cis</u>-3,4-dimethyl-3-hexene

11.24 (a) $Br-CH_2CH_2-Br$ (b) $CH_3CH_3$ (c) $CH_3CH_2-Br$
(d) $CH_3CH_2-OH$ (e) $Cl-CH_2CH_2-Cl$

11.26 (a) $CH_3CH{=}CH_2 + H_2 \xrightarrow{Pt} CH_3CH_2CH_3$

(b) $CH_3CH{=}CH_2 + HBr \longrightarrow CH_3-\underset{\displaystyle Br}{\underset{|}{C}}H-CH_3$

(c) $CH_3CH{=}CH_2 + Br_2 \longrightarrow CH_3-\underset{\displaystyle Br}{\underset{|}{C}}H-CH_2-Br$

(d) $CH_3CH{=}CH_2 + H_2O \xrightarrow[H_2SO_4]{} CH_3-\underset{\displaystyle OH}{\underset{|}{C}}H-CH_3$

11.28 The product actually formed in each case is written on the left side.

(a) $CH_3-\underset{\displaystyle Br}{\underset{|}{C}}H-CH_3$ $CH_3CH_2CH_2-Br$

11.28 continued

(b) 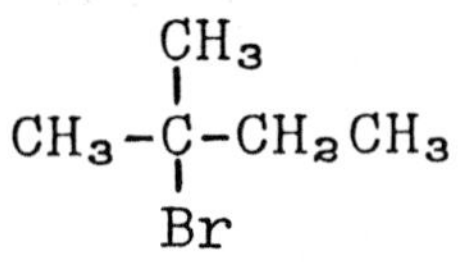 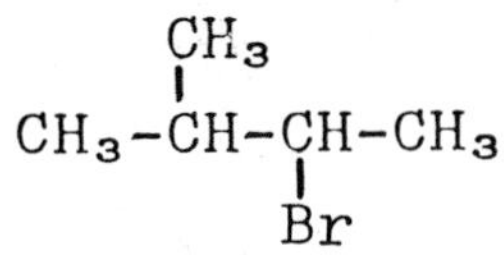

(c) 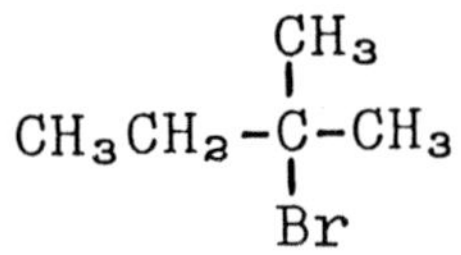 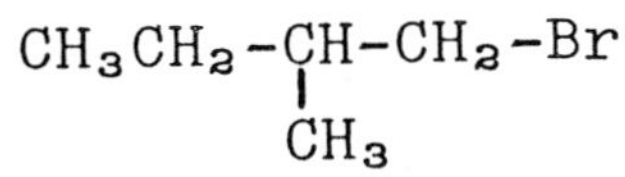

(d) 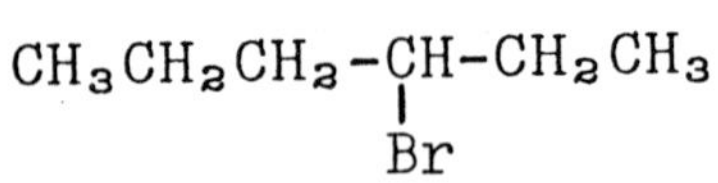 The only product in this case

(e) 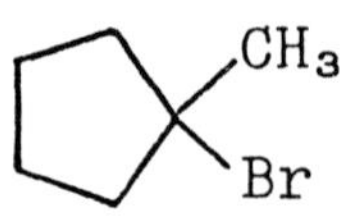 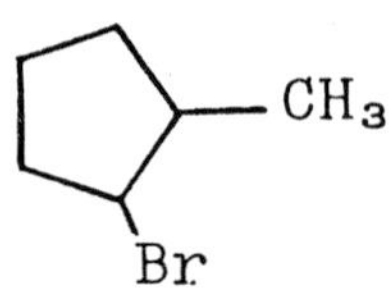

11.30 not by a direct addition, since HCl + propene gives 2-chloropropene

11.32 (a) 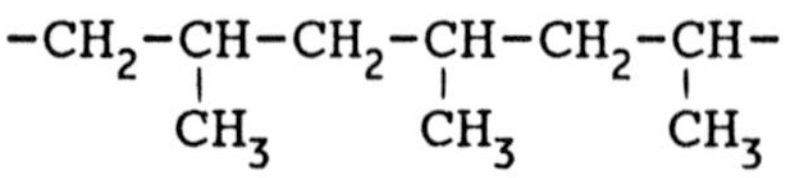

(b) 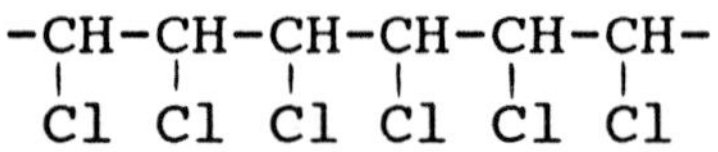

11.34 one sigma; two pi

11.36 (a) $H-C\equiv C-CH_3$ (b) $H-C\equiv C-CH_2CH_2CH_3$

(c) 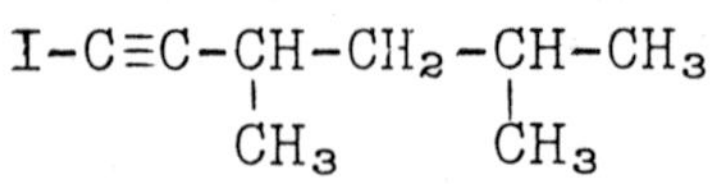

11.38 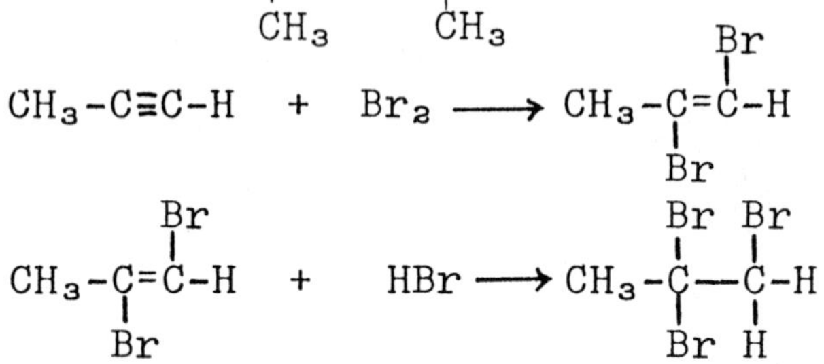

11.40 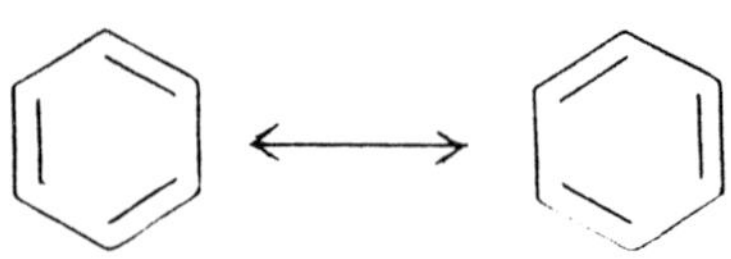

Neither represents the real benzene, which is an average of the two.

11.42 The entire molecule (all 12 atoms of $C_6H_4Cl_2$) lies in a single plane. No groups are above or below this plane.

11.44 (a)

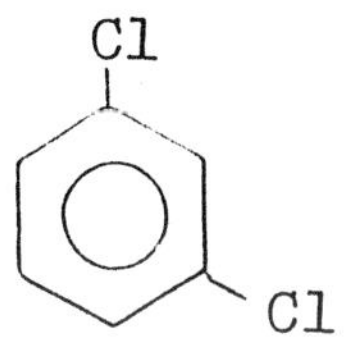

(b)

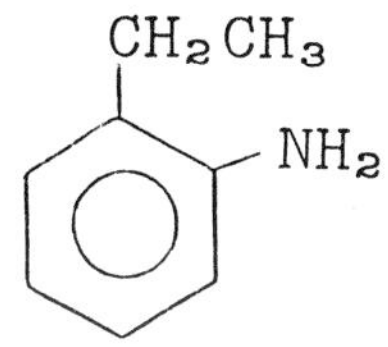

(c)

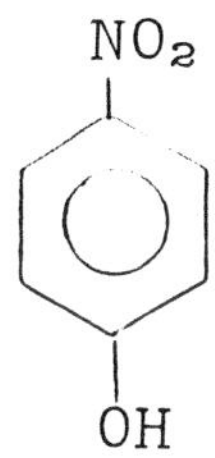

(d)

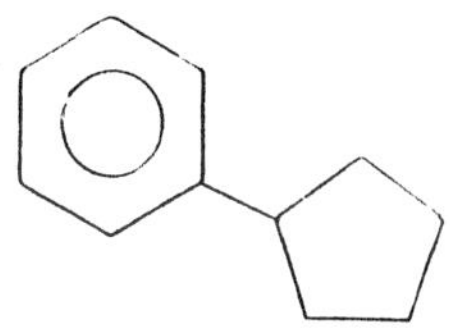

(e)

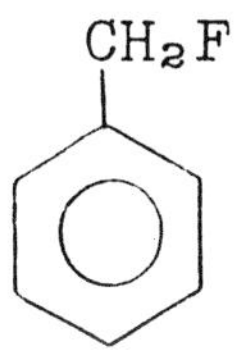

(f)

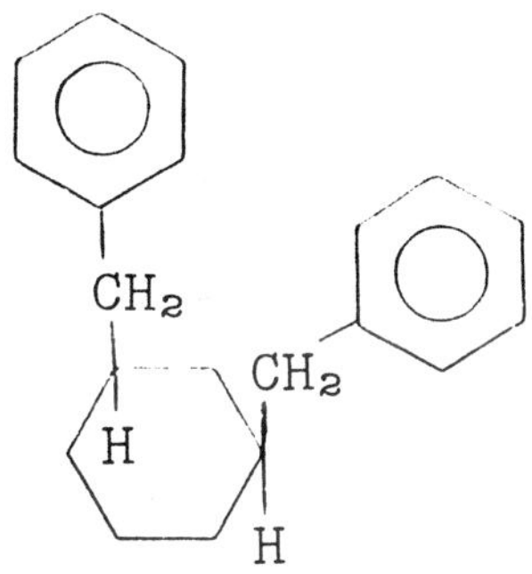

(g)

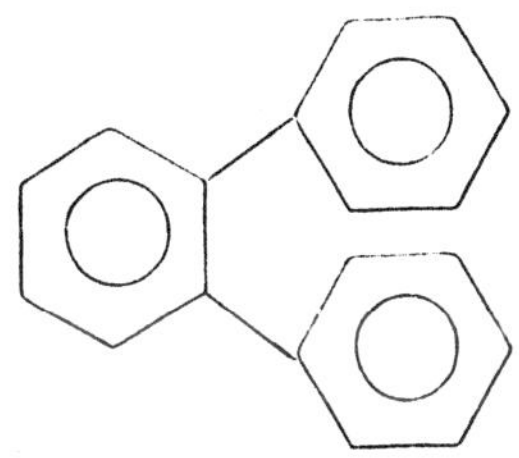

(h)

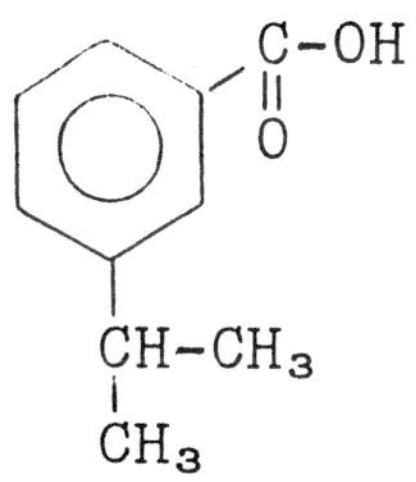

11.46 (a) 1,2,3-tribromobenzene (b) 2,3,4-tribromotoluene
(c) 3,5-dichlorophenol (d) 2,6-dinitrobenzoic acid
(e) 2-methyl-3-phenylpentane
(f) 1-bromo-1,2-diphenylethane (g) benzyl iodide

11.48 (a) $HNO_3$, $H_2SO_4$ (b) $Br_2$, Fe (c) first $Cl_2$, Fe; then $Br_2$, Fe, or the reverse (d) first $HNO_3$, $H_2SO_4$; then concentrated $H_2SO_4$, or the reverse

11.50 (a) no reaction (b)

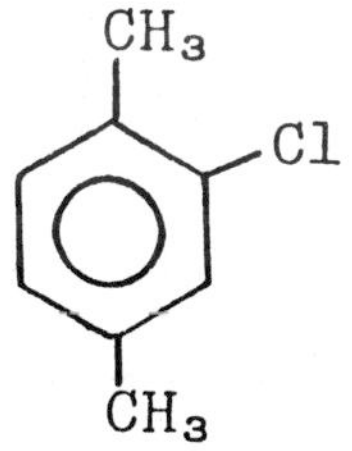

(c) $H-C\equiv C-H$

11.52

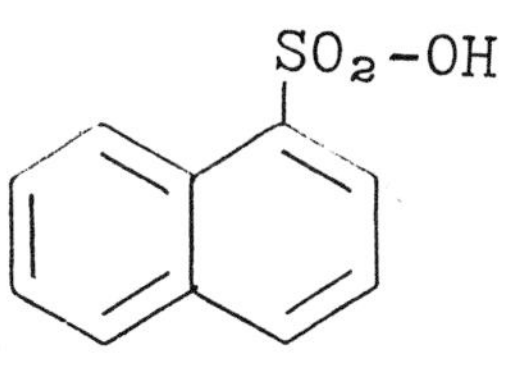

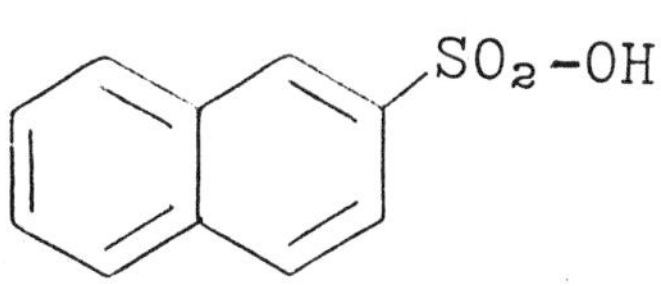

11.54

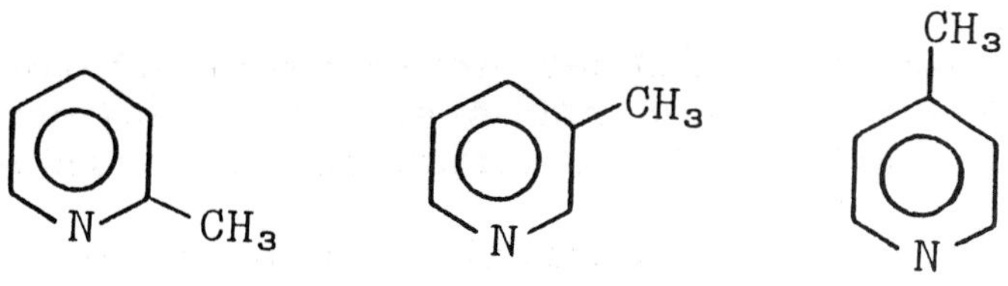

11.56

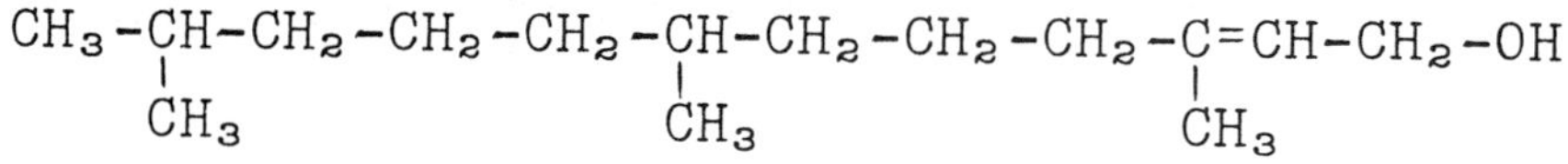

11.58 1.9 x $10^9$ molecules, or 1.9 billion molecules

11.60

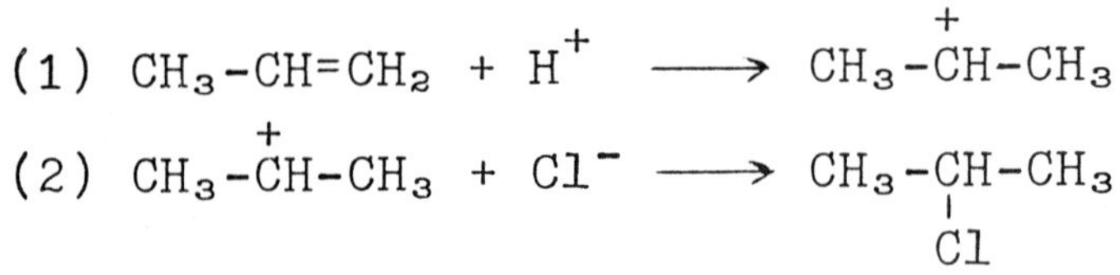

11.62

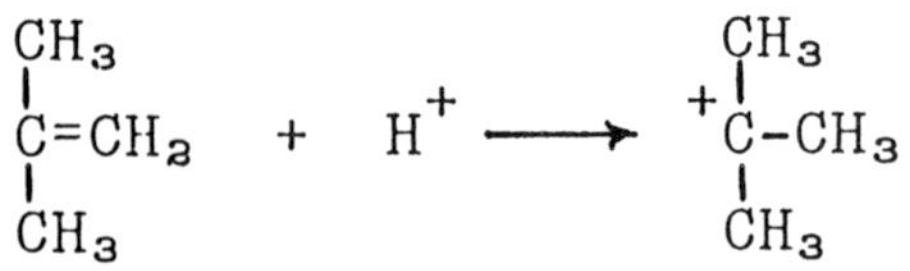

carbocation

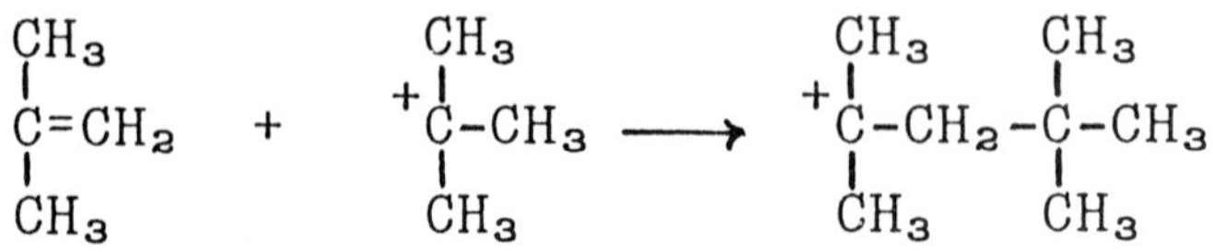

dimer

11.64 A compound that causes cancer. Condensed aromatic hydrocarbons, such as benzo(e)pyrene.

11.66 The atomic orbitals that make up the pi bond overlap less than those that make up the sigma bond.

11.68 (a) 2-Methyl-2-pentene does not have cis and trans isomers.

(b) The prefix meta may not be used if 3 or more groups are on a benzene ring.

(c) Since C forms only four bonds, there cannot be two methyl groups attached to C-2, which already has a single and a double bond.

(d) A lower number would be achieved by counting from the other end.

(e) A double bond takes precedence over a halogen.

(f) When a methyl group is on a benzene ring the parent name toluene shoud be used.

(g) and (h) There cannot be a bromine attached to the carbons indicated, which already have 4 bonds without it.

(i) There is a missing number. The position of the double bond is not given.

11.70 A triple bond requires a linear arrangement of four atoms: C-C≡C-C. If they were in a 6-membered ring, the bonds would have to bend this way 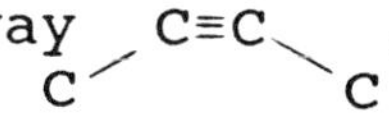, which is impossible.

11.72 (a) A double bond in a hydrocarbon must be given the lowest possible numbers, in this case 1 and 2.
(b) The correct name is 3,4-dimethylcyclohexene.

*Chapter 12. Alcohols, Phenols, Ethers, and Halides*

12.10 Primary: (a)(d)(h)(l) Secondary: (b)(c)(f)(g)(j)
Tertiary: (e)(i)(k)

12.12 See Problem 12.10 and its answer.

12.14 (a) 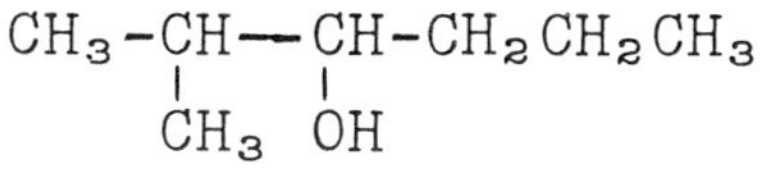 (b) 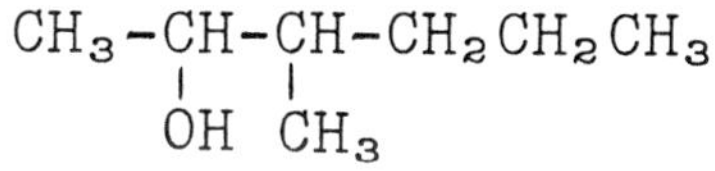

(c) 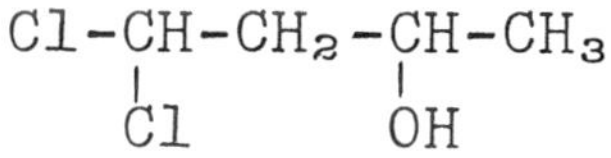 (d) 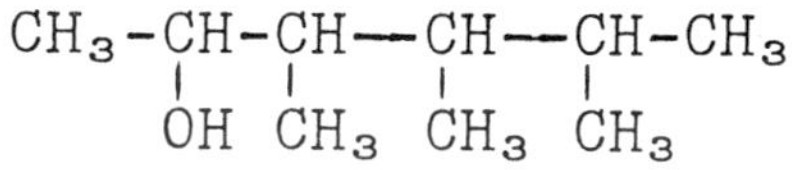

(e) 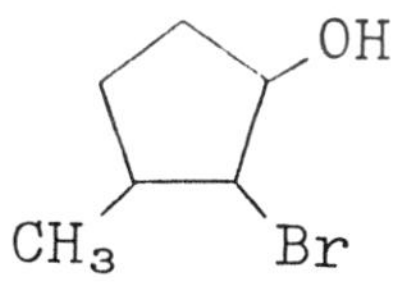

12.16 (a) $CH_3CH_2$-O-$CH_2CH_3$ (b) $CH_2$=$CH_2$ (c) $CH_3COOH$

12.18 (a) ethanol that has been made unfit to drink
(b) 100% alcohol
(c) a solution in which the solvent is ethanol

12.20 (a) phenol (b) <u>meta</u>-bromophenol (c) hydroquinone
(d) 3-chloro-5-nitrophenol

12.22 (a) picric acid (b) phenol (c) vanillin

12.24 (a) $CH_3$-O-$CH_3$ (b) $CH_2CH_3$-O-$CH_2CH_2CH_3$ (c) 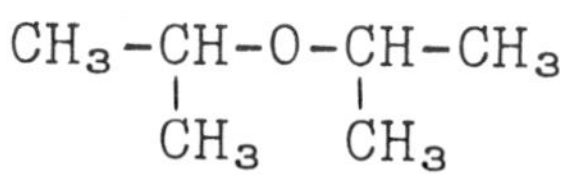

(d) 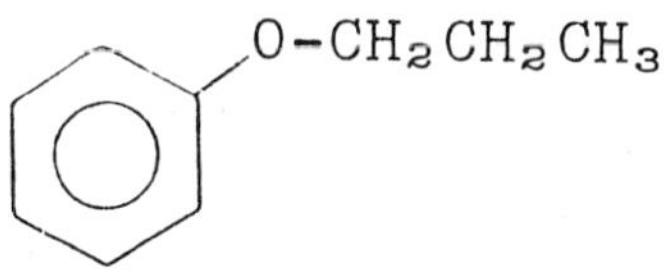 (e) 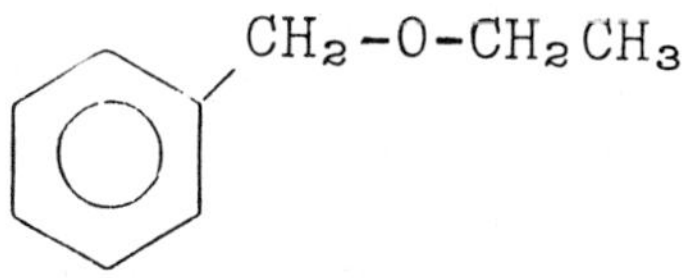

12.26 (a) (1)(2) (b) (1)(2)(5) (c) (2)(3)
(d) (4)(5) (e) (1)(4) (f) (2)(5)

12.28 (1) $CH_3CH_2CH_2CH_2-OH$ (2) $CH_3-\underset{OH}{\underset{|}{C}}H-CH_2CH_3$

(3) $CH_3-\underset{CH_3}{\underset{|}{C}}H-CH_2-OH$ (4) $CH_3-\overset{CH_3}{\overset{|}{\underset{CH_3}{\underset{|}{C}}}}-OH$

(5) $CH_3-O-CH_2CH_2CH_3$ (6) $CH_3-O-\underset{CH_3}{\underset{|}{C}}H-CH_3$

(7) $CH_3CH_2-O-CH_2CH_3$

12.30 (a) $CH_3CH_2COOH$ (b),(c) no reaction (d) $CH_3CH_2-\underset{O}{\underset{\|}{C}}-CH_2CH_3$

12.32 (a) $CH_3CH_2CH_2OH$ (b) $CH_3CH_2-\underset{OH}{\underset{|}{C}}H-CH_3$ (c) $CH_3CH_2CH_2CH_2OH$

12.34 Ethylene glycol, with two OH groups, participates in more extensive hydrogen bonding than 1-propanol.

12.36 (a) thiols (b) ethers (c) phenols (d) alcohols

12.38 (a) $CH_3CH_2-S-S-CH_2CH_3$ (b) $CH_3-SH$

12.40 Acidic: phenols, thiols Neutral: alcohols, ethers, halides, thioethers, disulfides

12.42 The $CO_2$ in the champagne causes the alcohol to get into the bloodstream faster.

12.44 by distillation

12.46 Alcohol impairs sensory perception and slows reflexes.

12.48 It relaxes the smooth muscles of the blood vessels and dilates the arteries, allowing more blood to reach the heart.

12.50 In surgery, they kill bacteria that would otherwise infect wounds.

12.52 Advantages: easy to administer and safe
Disadvantages: highly flammable, slow-acting, causes nausea, irritates the respiratory tract

12.54 A biodegradable substance can be decomposed by natural processes in the environment.

12.56 Low iodine levels in the soil mean low iodine levels in the food; hence there is not enough iodine available for the body to manufacture the hormone.

12.58 (a) $CH_3CH_2CH_2-\underset{O}{\underset{\|}{C}}-H$ and $CH_3CH_2CH_2-\underset{O}{\underset{\|}{C}}-OH$
(b) the second one (the carboxylic acid)
(c) by distilling the reaction mixture before the aldehyde is further oxidized

12.60 Molecules of 1-butanol form hydrogen bonds with each other. Molecules of methyl propyl ether cannot do this.

12.62 (b)(c)(a)

12.64 $CH_3SH$ is not capable of forming hydrogen bonds with other molecules of itself; $CH_3OH$ does form such bonds.

12.66 Menthol is an alcohol, hence neutral; thymol is a phenol; hence acidic.

## *Chapter 13. Aldehydes and Ketones*

13.8 Aldehydes have the formula $R{-}\underset{\displaystyle O}{\overset{}{\underset{\|}{C}}}{-}H$ or $H{-}\underset{\displaystyle O}{\underset{\|}{C}}{-}H$, where R = an alkyl group or aromatic ring. Ketones have the formula $R{-}\underset{\displaystyle O}{\underset{\|}{C}}{-}R'$.

13.10

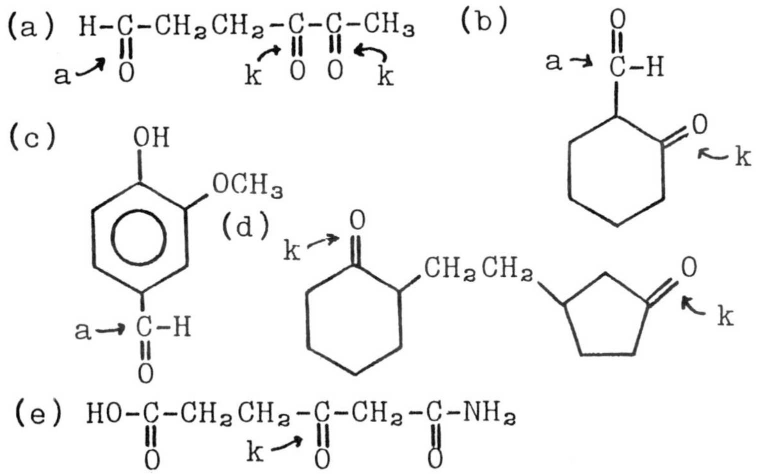

13.12 (a) benzaldehyde (b) acetone (c) methyl n-propyl ketone (d) formaldehyde (e) acetaldehyde (f) acetophenone

13.14

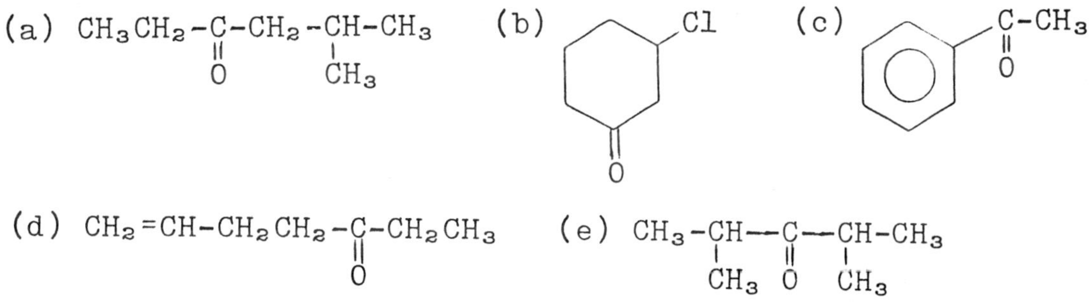

13.16 4-Heptanone is much bigger, and water cannot solvate the two nonpolar ends of the molecule.

13.18

13.20 (a) $CH_3CH_2\text{-}\overset{\displaystyle O}{\overset{\|}{C}}\text{-}OH$ (b) no reaction (c) 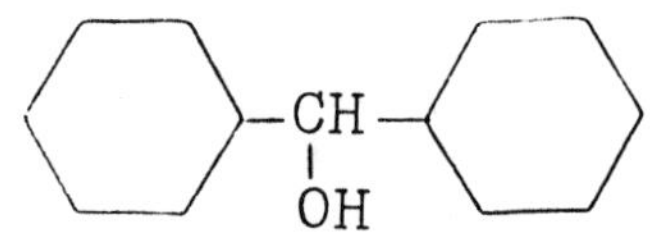

(d) $CH_3CH_2CH_2CH_2\text{-}OH$

13.22 $R\text{-}\underset{\displaystyle O}{\underset{\|}{C}}\text{-}H + 2Cu^{2+} + 5OH^- \longrightarrow R\text{-}\underset{\displaystyle O}{\underset{\|}{C}}\text{-}O^- + Cu_2O + 3H_2O$

13.24 (b)(c)(e)

13.26 (a) $CH_3CH_2\text{-}\underset{\displaystyle OCH_2CH_3}{\underset{|}{CH}}\text{-}OCH_2CH_3$ (b) 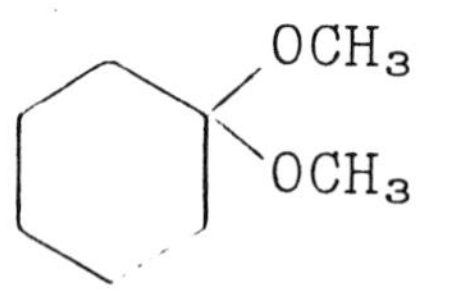 (c) 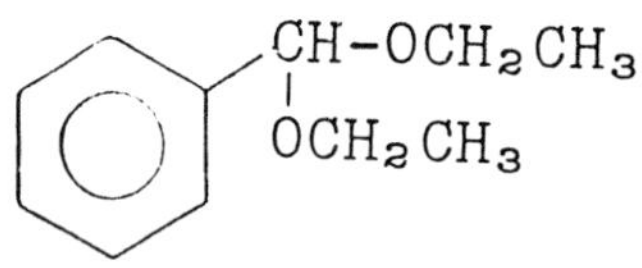

(d) $HO\text{-}CH_2CH_2CH_2\text{-}\underset{\displaystyle O}{\underset{\|}{C}}\text{-}H + CH_3OH$

13.28 $CH_3\text{-}\overset{\displaystyle CN}{\overset{|}{\underset{\displaystyle OH}{\underset{|}{C}}}}\text{-}H$

13.30 Ketals: (b)(c)(e) Hemiketals: (a)(f) Neither: (d)(g)

13.32 (a) $CH_3CH_2\text{-}\underset{\displaystyle OH}{\underset{|}{CH}}\text{-}OCH_2CH_3$ $CH_3CH_2\text{-}\underset{\displaystyle OCH_2CH_3}{\underset{|}{CH}}\text{-}OCH_2CH_3$

(b) 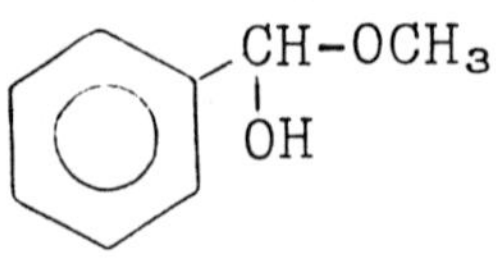 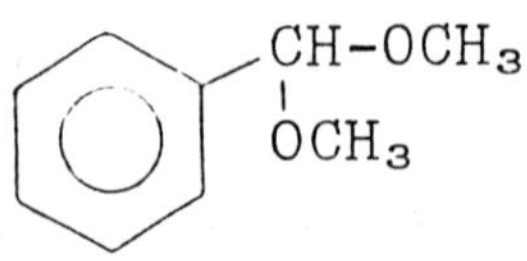

(c) 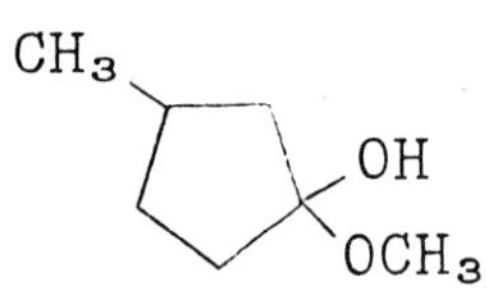 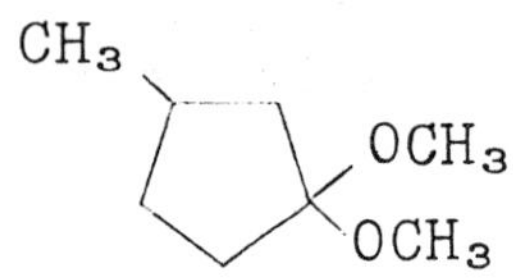

(d) 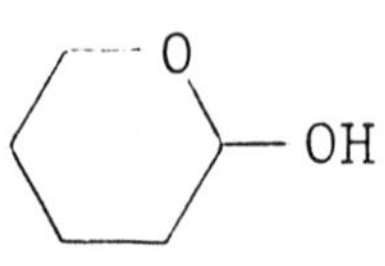

13.34 (a) $CH_3-C(=O)-CH_3$ (b) 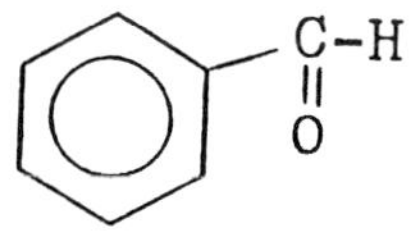 (c) $H-C(=O)-H$

(d) none (e) 

13.36 (a)(b)(d)(f)

13.38 (a) $CH_3-C(=O)-H$ (b) $CH_3CH_2-C(=O)-CH_3$ (c) $H-C(=O)-CH_2CH_2-C(=O)-H$

(d) 

13.40 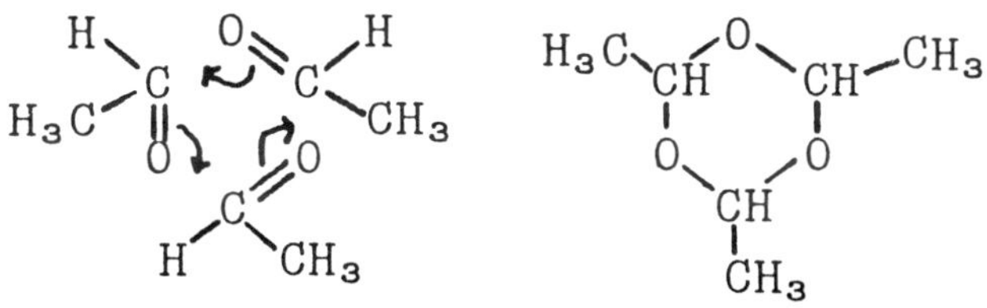

13.42 $Cl_3C-CH(OH)(OCH_2CH_3)$

13.44 $CH_3-C(=O)-H + H^- \longrightarrow CH_3-CH_2-O^- \xrightarrow{H_2O} CH_3-CH_2-OH$

13.46 plastics, adhesives, coatings
13.48 yes
13.50 ketone, cyclic alkene
13.52 It tautomerizes to form the more stable compound acetone: $CH_3-C(=O)-CH_3$

13.54 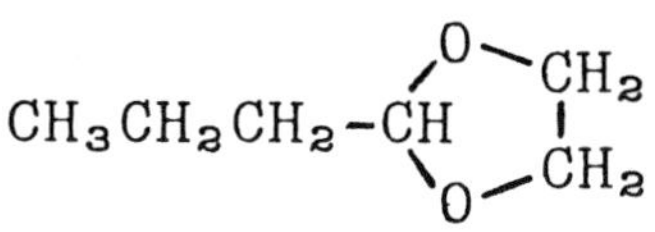 This is an acetal

13.56 (a) 2-bromo-1-chloro-3-hexanone
(b) 1-phenyl-2-butanone
(c) 3,4-dibromobenzaldehyde
(d) pent-1-en-3-one
(e) vanillin or 4-hydroxy-3-methoxybenzaldehyde
(f) 2,4-dimethylcyclopentanone

13.58 The enol form is stabilized by internal hydrogen bonding:

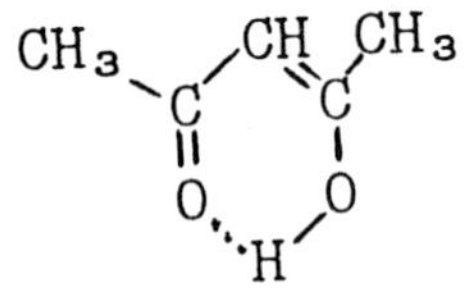

13.60 Ketones have pleasant odors. Short-chain aldehydes have unpleasant odors; long-chain ones pleasant.

13.62 Formaldehyde (b.p. -21°C) is a gas at room temperature. Acetaldehyde (b.p. 21°C) is a gas in the summer and a low-boiling liquid in the winter.

*Chapter 14. Carboxylic Acids and Esters*

14.10 (a) pentanoic acid (b) 2-bromo-4,5-dimethylhexanoic acid (c) 2,3-diphenylpropanoic acid (d) 4,6-dibromo-3,5-dichloro-2-methylhexanoic acid

14.12 (a) acetic acid (b) caprylic acid (c) glutaric acid (d) benzoic acid (e) maleic acid (f) palmitic acid

14.14 (a) hexanoic acid (b) decanoic acid
(c) 2,2-dimethylpropanoic acid

14.16 (a) unbranched carboxylic acids from $C_3$ to $C_{20}$
(b) the esters of most of them are found in fats

14.18

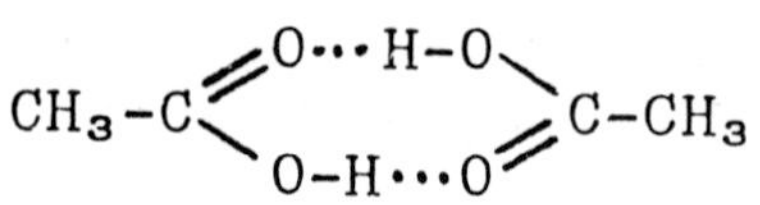

14.20 (c) (a) (b)

14.22 The -COOH group is responsible for the solubility. The $CH_3-(CH_2)_4-$ chain prevents greater solubility.

14.24 $CH_3CH_2OH + O_2 \longrightarrow CH_3COOH + H_2O$

14.26 (a) potassium acetate (b) sodium caprylate
(c) dipotassium succinate (d) lithium benzoate

14.28 It turns into the acid $CH_3COOH$.

14.30 Add aqueous NaOH solution. Capric acid will be converted to sodium caprate, which is soluble in water. 1-Decanone stays out of solution.

14.32 The alcohol or phenol part of each carboxylic ester is circled. (b), (d), and (g) are not esters.

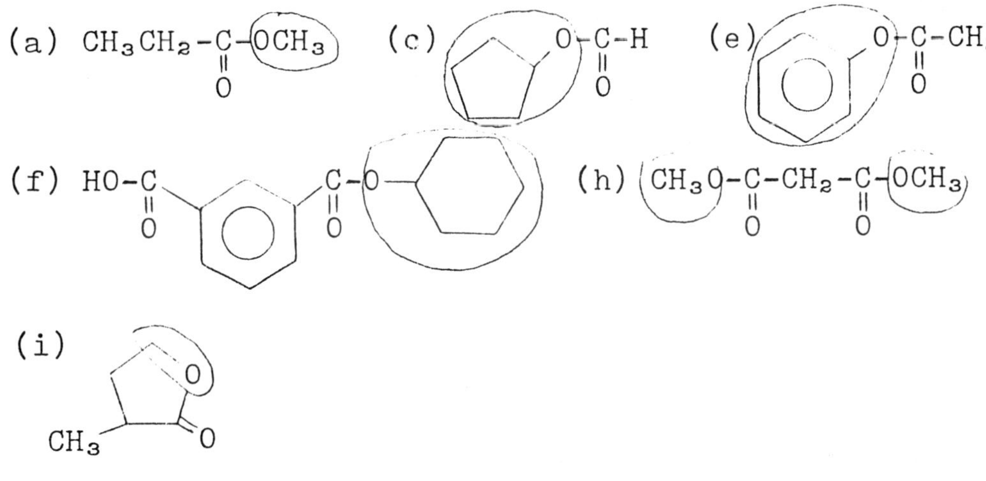


14.34 (a) $H-C(=O)-OCH_2CH_3$ (b) $CH_3CH_2CH_2-C(=O)-OCH_3$

(c) $CH_3CH_2CH_2CH_2-C(=O)-OCH_2CH_2CH_2CH_3$ (d) $CH_3(CH_2)_6-C(=O)-CH(CH_3)-CH_3$

(e) $CH_3O-C(=O)-CH_2-C(=O)-OCH_3$ (f)

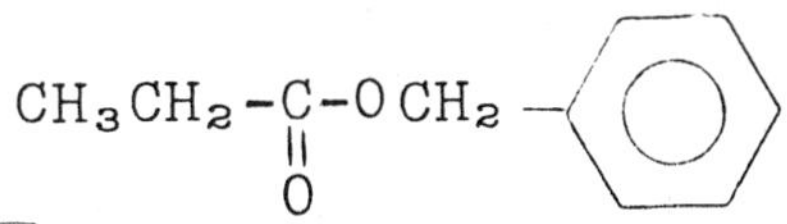


(g)

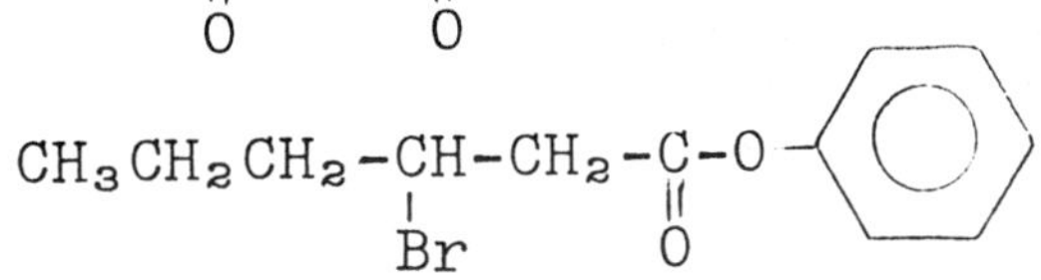


14.36 About the same solubility, because both form hydrogen bonds with water equally well, and they have the same molecular weight.

14.38 carboxylic acids: (b)(j)(k)
carboxylic esters: (a)(c)(g)(k)
acyl chloride: (d)
anhydrides: (f)(i)
none of these: (e)(h)

14.40 (a) $CH_3CH_2CH_2COOCH_2CH_3 + H_2O \xrightarrow{H^+} CH_3CH_2CH_2COOH + CH_3CH_2OH$

(b) $CH_3CH_2CH_2COOCH_2CH_3 + H_2O \xrightarrow{OH^-} CH_3CH_2CH_2COO^- + CH_3CH_2OH$

14.42 (a) $CH_3COC^- + CH_3OH$

(b) $CH_3COOH + CH_3OH$

14.42 continued

(c) $CH_3CH_2COO^-$ + 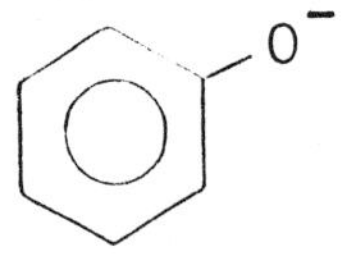

(d) 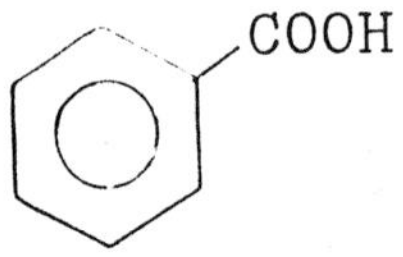 + $CH_3CH_2OH$

(e) $^-O-C(=O)-CH_2-C(=O)-O^-$ + $CH_3OH$

(f) $CH_3CH_2COO^-$ + $HO-CH_2CH_2-OH$

(g) $CH_3(CH_2)_8COO^-$ + $CH_3-CH(CH_3)-CH_2OH$

14.44 (a)

$HO-P(=O)(OH)-OCHCH_2CH_3$ $\qquad$ $CH_3CH_2O-P(=O)(OH)-OCH_2CH_3$

$CH_3CH_2O-P(=O)(OCH_2CH_3)-OCH_2CH_3$

(b) two, one, none, respectively

14.46

$$HO-P(=O)(OH)-OH + H-O-P(=O)(OH)-OH + H-O-P(=O)(OH)-OH \longrightarrow$$

$$HO-P(=O)(OH)-O-P(=O)(OH)-O-P(=O)(OH)-OH \quad + \text{ two } H_2O$$

14.48 (a)

$$^-O-P(=O)(O^-)-O-P(=O)(O^-)-O-P(=O)(O^-)-OCH_3 + H_2O \xrightarrow{\text{enzyme}} {}^-O-P(=O)(O^-)-OH + HO-P(=O)(O^-)-O-P(=O)(O^-)-OCH$$

(b)

$$^-O-P(=O)(O^-)-O-P(=O)(O^-)-OCH_3 + H_2O \xrightarrow{\text{enzyme}} {}^-O-P(=O)(O^-)-OH + HO-P(=O)(O^-)-OCH_3$$

14.50 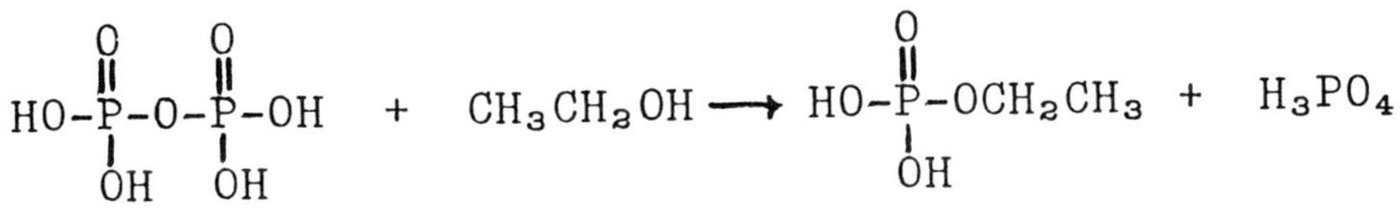

14.52 a tricarboxylic acid

14.54 $(CH_3-\underset{\overset{\|}{O}}{C}-O^-)_3 \quad Al^{3+}$

14.56 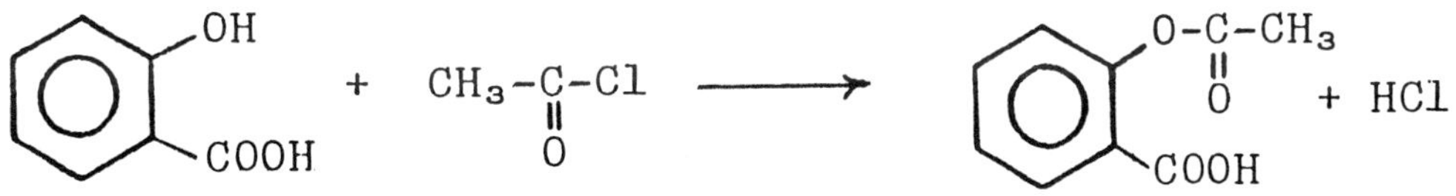

14.58 See Box 14E.

14.60 They are made of a polyester that hydrolyzes in the body to its components, lactic and glycolic acids. These are soluble in body fluids.

14.62 Caprylic acid forms the salt $CH_3(CH_2)_6COO^-$, which is soluble in water. Since caprylaldehyde is not acidic, it is not converted to a salt.

14.64 Less soluble. Ionic compounds are generally soluble in water; high-molecular weight acids are not.

14.66 (a) $CH_3-\underset{\overset{\|}{O}}{C}-O-CH_2CH_3$ (b) $CH_3-\underset{\overset{\|}{O}}{C}-O-\underset{\overset{\|}{O}}{C}-CH_3$ (c) $CH_3-\underset{\overset{\|}{O}}{C}-Cl$

14.68 (a) 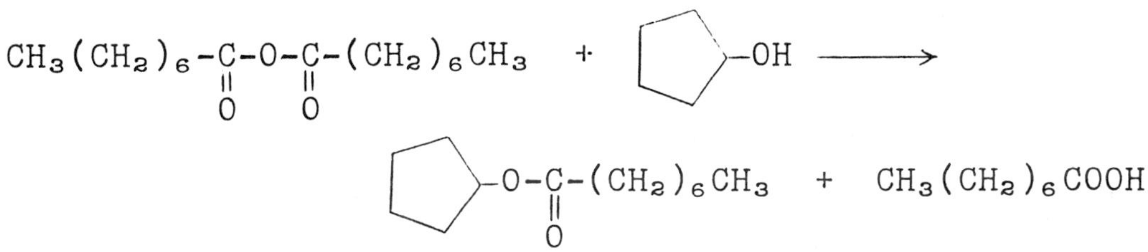

(b) 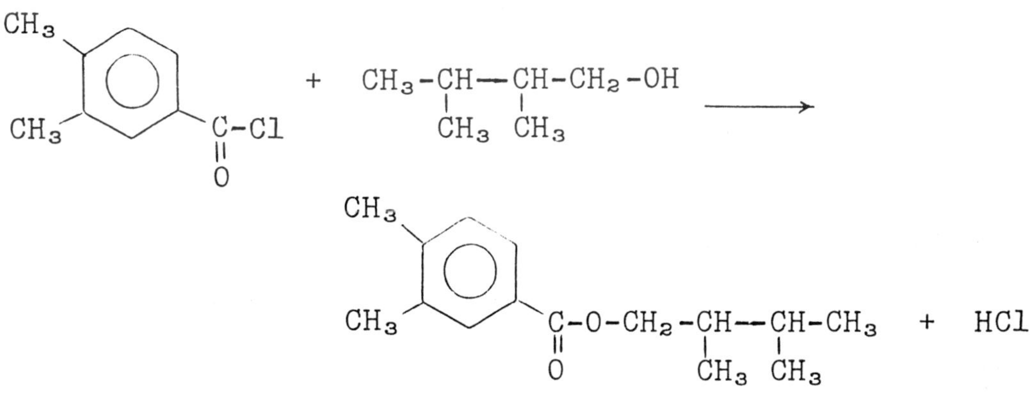

*Chapter 15. Amines and Amides*

15.12 ammonia

15.14 Primary: (a)(c) Secondary: (d)(f) Tertiary: (b)(e)

15.16 Primary: (a),(d),(h),(i),(l),(m)
Secondary: (b),(c),(j)
Tertiary: (e),(f),(g),(k)

15.18 (a) $CH_3NH_2$ (b) $CH_3CH_2CH_2CH_2\text{-NH-}CH_2CH_2CH_2CH_3$

(c) $CH_3CH_2\text{-N-}CH_2CH_3$ with $CH_2CH_3$ on N

(d) 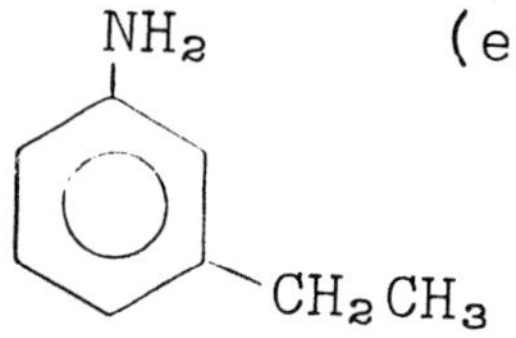

(e) 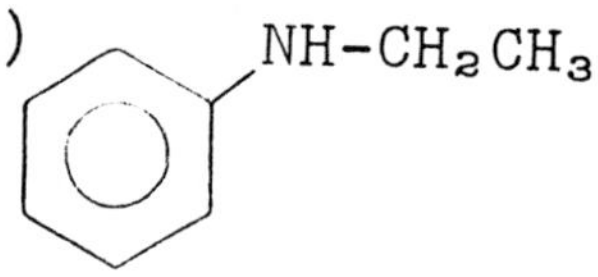

(f) $CH_3\text{-CH-}CH_2\text{-NH-CH-}CH_2CH_3$ with $CH_3$ on each CH

(g) $NH_2(CH_2)_4NH_2$

(h) $NH_2\text{-}CH_2CH_2\text{-CH-}CH_2CH_2\text{-}NH_2$ with $NH_2$ on CH

15.20 Like alcohols, they form hydrogen bonds to water.

15.22 (a) $CH_3CH_2CH_2NH_3^+\ Cl^-$ (b) 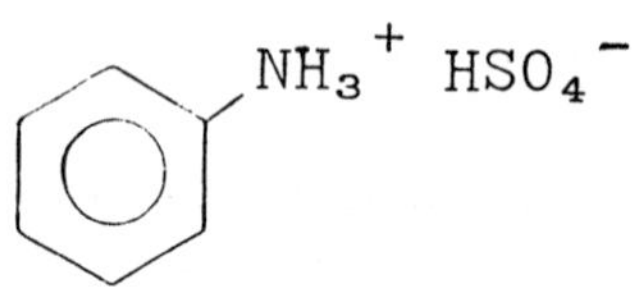

(c) 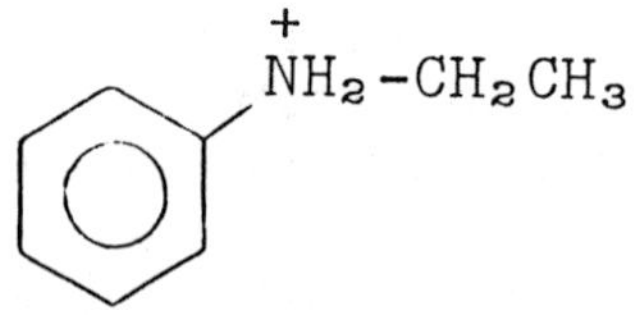 $H_2PO_4^-$

(d) $CH_3CH_2\text{-}\overset{+}{N}H\text{-}CH_3$ with $CH_3$ on N, $CH_3COO^-$

(e) $CH_3CH_2\text{-}\overset{+}{N}H_2\text{-}CH_2CH_3$ 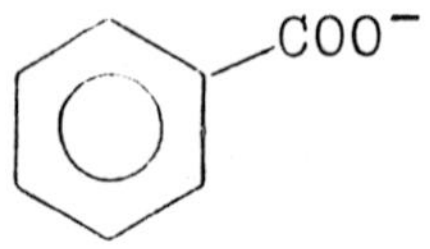

15.24 (a) $CH_3CH_2CH_2NH_3^+\ Cl^-$ (b) $CH_3CH_2CH_2CH_2\text{-}\overset{+}{N}\text{-}CH_3$ with two $CH_3$ on N, $OH^-$

(c) 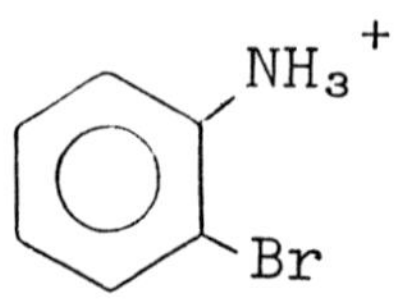 $CH_3COO^-$

(d) $CH_3\text{-CH-}CH_2\text{-}\overset{+}{N}H_2\text{-}CH_2CH_3$ with $CH_3$ on CH, $Br^-$

15.26 isopropylammonium bromide

$CH_3-\underset{\displaystyle CH_3}{CH}-NH_3^+\ Br^-$ (with $CH_3$ attached above the central CH)

15.28 Add HCl.

15.30 (a) 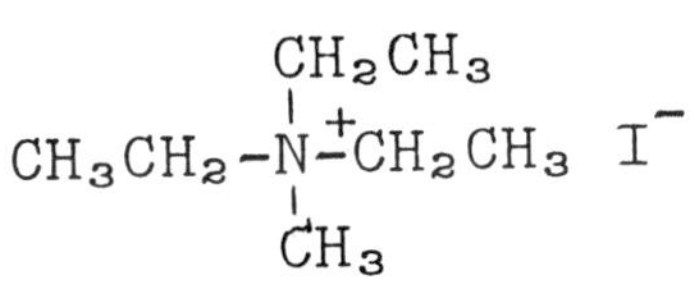 (b) 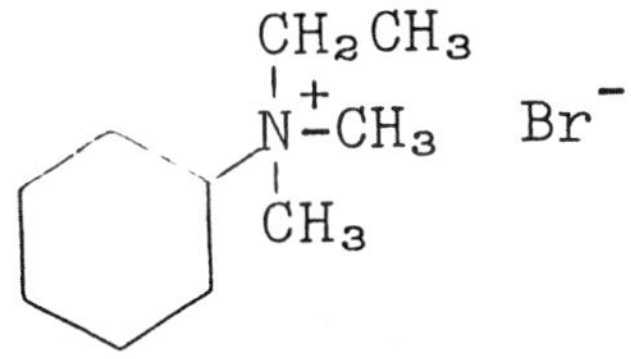

(c) 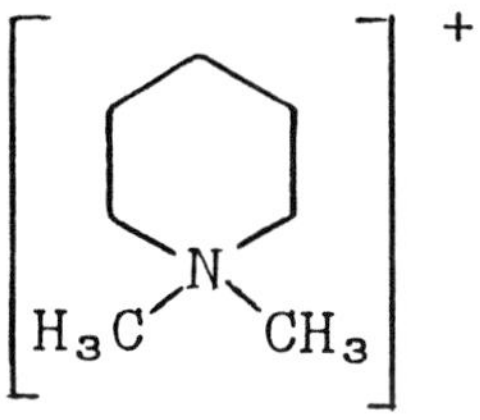 $Cl^-$

15.32 Add chloromethane to ethyldimethylamine or add chloroethane to trimethylamine.

15.34 Amine: (b)(j) Amide: (c)(e)(f)(g)(i)(k)
Both: (d)(h) Neither: (a)

15.36 (a) formamide (b) N-ethylbenzamide
(c) N,N-dimethylacetamide (d) N-phenylbutyramide
(e) N-methylpalmitamide
(f) ortho-bromo-N,N-diethylbenzamide

15.38 No; add the amine to an acyl halide or an anhydride or to a carboxylic acid in the presence of DCC.

15.40 Water would react with the DCC, turning it into DHU. Without the DCC the amide would not form.

15.42 (a) and (c) $H_2O$ + $OH^-$ (b) and (d) $H_2O$ + $H^+$

15.44 a basic compound obtained from a plant

15.46 (a) phenol, secondary alcohol, secondary amine
(b) carboxylic ester, primary alcohol, tertiary amine
(c) double bond, secondary alcohol, ether, phenol, tertiary amine
(d) tertiary amine, carboxylic ester

15.48 See Box 15A.

15.50 The salt is more soluble in body fluids.

15.52 The nitrite ion becomes nitrous acid in the acidic environment of the stomach. This reacts with secondary amines to form N-nitrosamines.

15.54 Acetaminophen and ibuprofen do not cause aspirin's side effects, and are still effective pain relievers. Both also reduce fever, but acetaminophen does not reduce swelling.

15.56 $-NH-\underset{\displaystyle \overset{\|}{O}}{C}-(CH_2)_4-\underset{\displaystyle \overset{\|}{O}}{C}-NH-(CH_2)_4-NH-\underset{\displaystyle \overset{\|}{O}}{C}-$

15.58 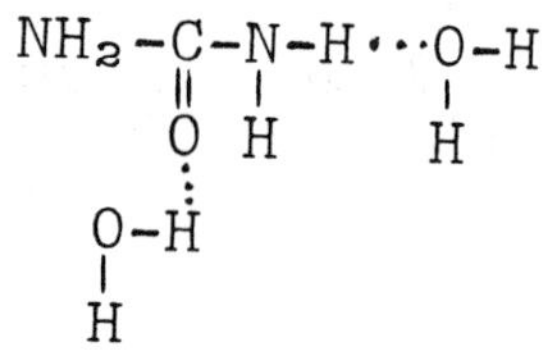

15.60 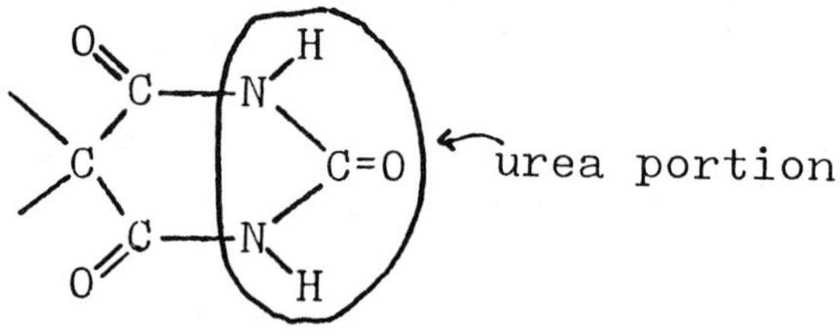


15.62 no

15.64 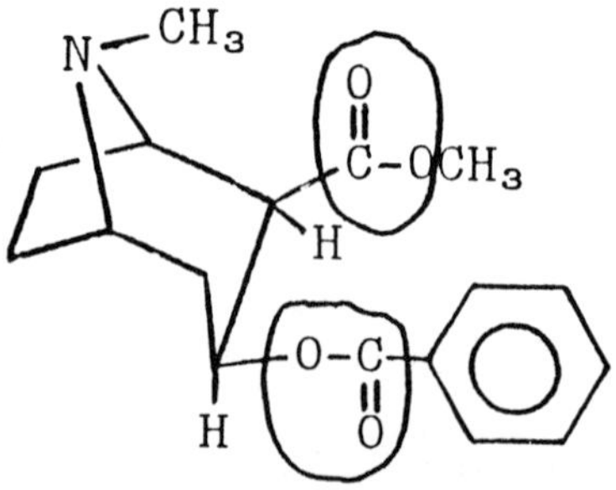

the carboxylic ester groups are circled

15.66 Add the mixture to a solution of HCl in water. The amine will be converted to its salt, which will dissolve in the water; the alcohol will remain as a separate layer that can be removed in a separatory funnel. The amine can be recovered by making the water layer basic.

15.68 (a) It is an ionic compound (a salt).
(b) It is a quaternary ammonium salt.

15.70 (a) and (b) are purely synthetic compounds, made entirely in laboratories. (c) is not found in nature; it is made from an alkaloid. (d) has no nitrogen and is not basic.

15.72 See the second part of Section 15.4.

15.74 The effect (enlarged pupils) lasts for hours.

15.76 by adding a base, $OH^-$

*Chapter 16. Carbohydrates*

16.4 (a) about 75% (b) In humans less than 1% of body weight is carbohydrates.

16.6 (a) isomers that are nonsuperimposable mirror images of each other
(b) stereoisomers that are not enantiomers
(c) light that vibrates in only one plane
(d) a mixture of equal amounts of two enantiomers

16.8 82.5°C, 218.5°C, 1.4 g/100 mL, 0.9040 g/mL, -24.1°

16.10 (1) They rotate the plane of polarized light in opposite directions. (2) They react at different rates with chiral substances.

16.12 (a) 2 (b) no stereoisomerism (c) 4 (d) 4 (e) 4 (f) 8 (g) 4

16.14 (a) polyhydroxy aldehydes or ketones, or compounds that will yield them on hydrolysis (b) simple sugars of the formula $C_nH_{2n}O_n$, where $n = 3$ to 8

16.16 No. They differ in configuration only at carbon 1. Enantiomers differ at all carbons.

16.18 (a) -112.2° (b) It changes to -52.7°.

16.20

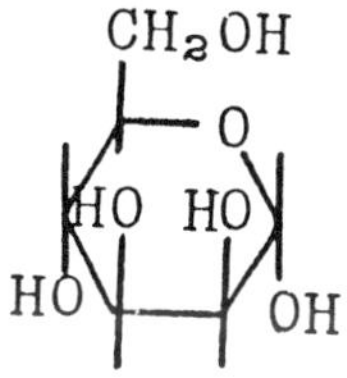

α isomer

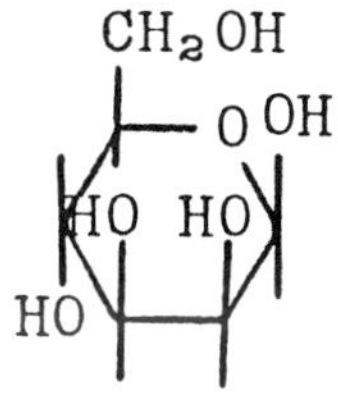

β isomer

16.22 carbon 6

16.24 They have numerous alcohol groups that can form hydrogen bonds with water.

16.26 (a)(b) See Sec. 16.6. (c)(d) See Sec. 16.8.

16.28 (a) The (-) means that it rotates the plane of polarized light to the left. (b) The common name is levulose; laevo means left in Latin.

16.30 sorbitol, also called glucitol

16.32 (a)

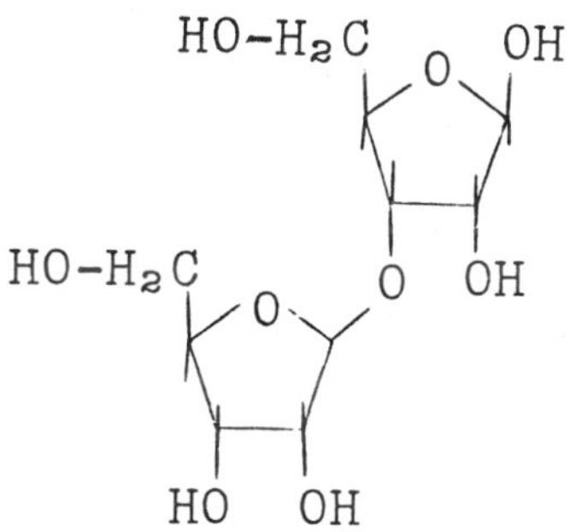

(b) yes

16.34 (a)

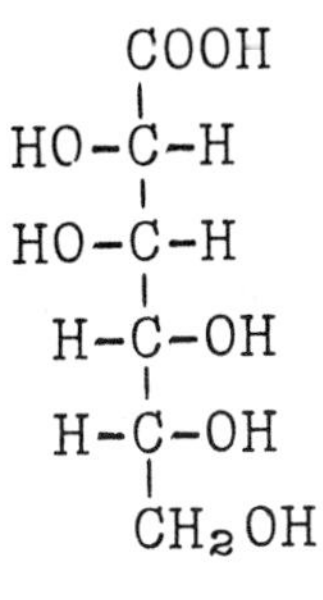

(b)

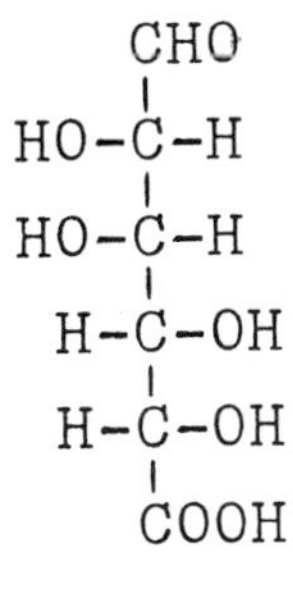

16.36 (a) yes

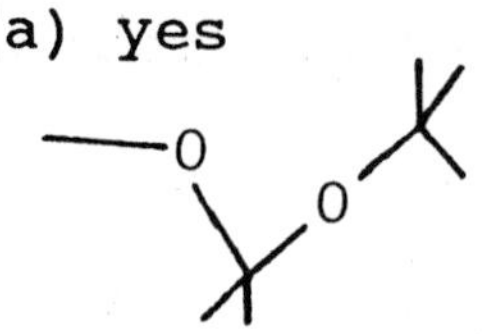

(b) yes

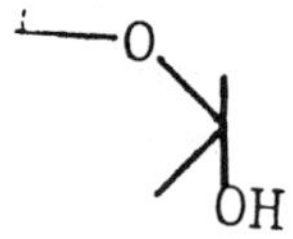

(c) yes

(d) This compound has a β-(1⟶4) glycosidic linkage. Maltose has an α-(1⟶4) glycosidic linkage.

16.38 The amylose forms helixes inside of which the $I_2$ molecules assemble in a long polyiodine chain.

16.40

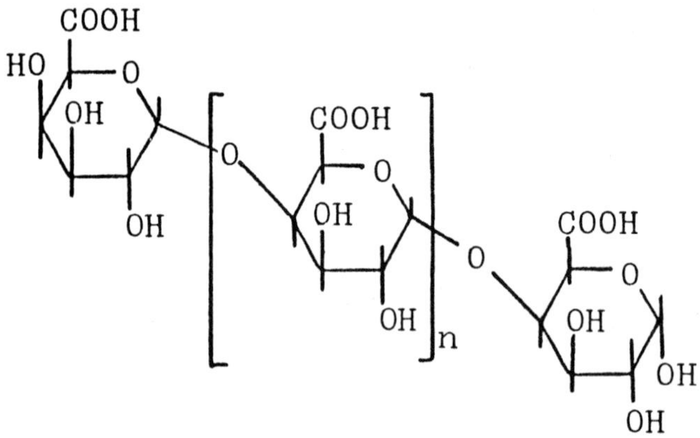

16.42 It decreases; the fluid becomes more watery, less viscous; hence it provides less lubrication.

16.44 Because in most cases nature makes only one of the possible stereoisomers of a given structure, and the others won't fit.

16.46 temperature and wavelength

16.48 (1) It is the meso isomer. (2) It is a racemic mixture of the 2 other isomers.

16.50 No; The OH on C-1 is in the axial position.

16.52

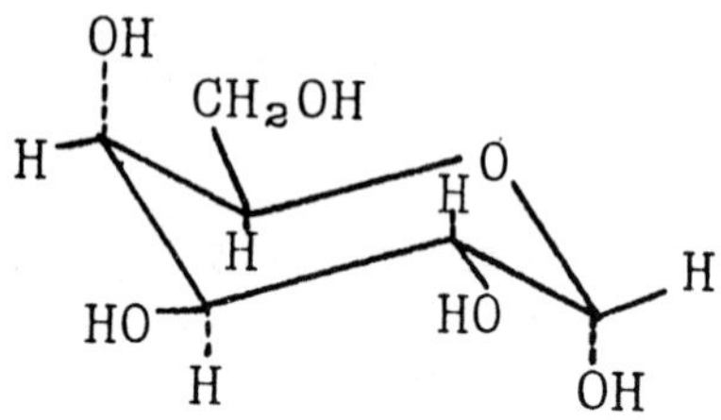

The OH groups on C-1 and C-4 are axial

16.54 (a) a protein manufactured in the body to combat foreign substances (antigens) (b) It clumps (aggregates) it.

16.56 B-type blood has B antigen. A person with A-type blood has antibodies against B antigen. If a transfusion is given, the antibodies will clump the transfused red blood cells and may cause death.

16.58 See Box 16E.

16.60 They are soluble in water and have good thickening and gelling properties.

16.62 See Box 16G.

16.64 No. Among organic compounds only amines are basic. Thalidomide is not an amine.

16.66 (a) maltose (b) amylose, amylopectin, starch

16.68 (b)(d)

16.70 The structure is identical to that of cellulose except that the -OH at the 2 position of each ring is replaced by $-NH-\overset{\overset{\displaystyle O}{\|}}{C}-CH_3$

16.72 D-glucose

16.74

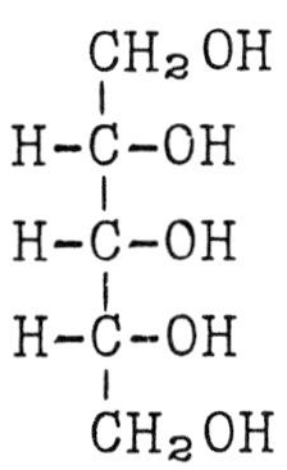

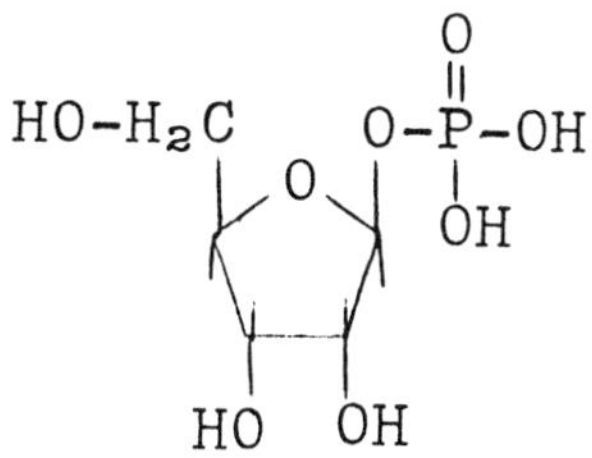

16.76 No; the ring is locked.

16.78 An equilbrium is being established among an open-chain and two cyclic ($\alpha$ and $\beta$) forms (mutarotation).

## *Chapter 17. Lipids*

17.2 Hydophobic means water hating. It is important because if the body did not have such molecules there could be no structure since the water would dissolve everything.

17.4 Because they are synthesized entirely from acetic acid units, two carbons at a time.

17.6 See Sec. 17.2.

17.8 Add $ZnCl_2$ to a solution of the potassium undecylenate.

17.10 Among similar compounds melting points increase with increasing molecular weight.

17.12 It is mostly made up of short chain fatty acids: $C_{12}$ and $C_{14}$.

17.14 5 moles.

17.16 (a) 15 (b) 10 (c) 5

17.18 The cis double bonds in the unsaturated fatty acids prevent the regular packing of the long aliphatic portions of the lipid bilayers, making them fluid-like.

17.20

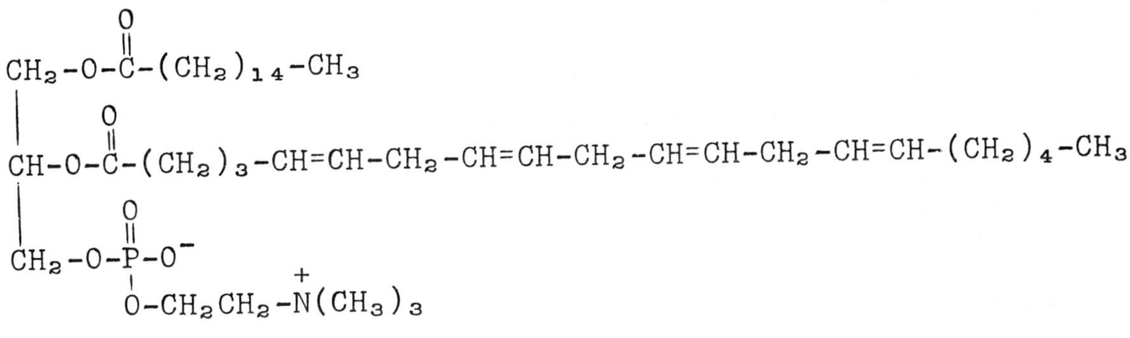

17.22

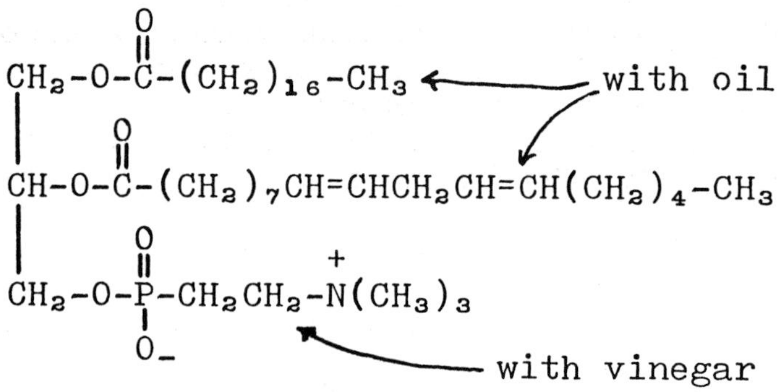


17.24 Fluidity. Cholesterol will not fit into the regular zig-zag packing of the fatty acid chains. Therefore, it disturbs the crystalline (high-melting) portions in a manner similar to the cis double bonds of unsaturated fatty acids.

17.26 (a) chiral carbons are circled.

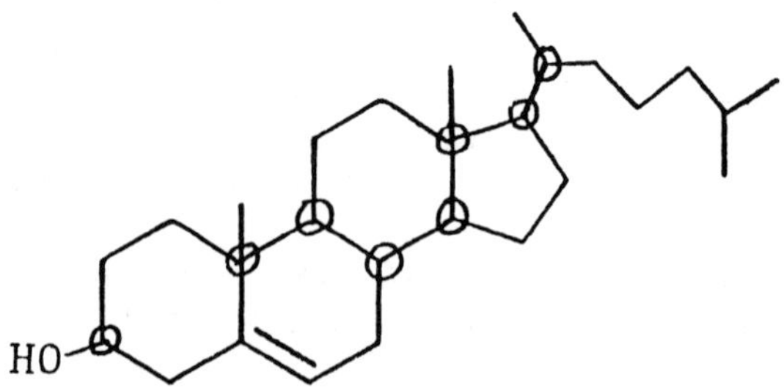


(b) $2^8 = 256$ (c) only 1

17.28

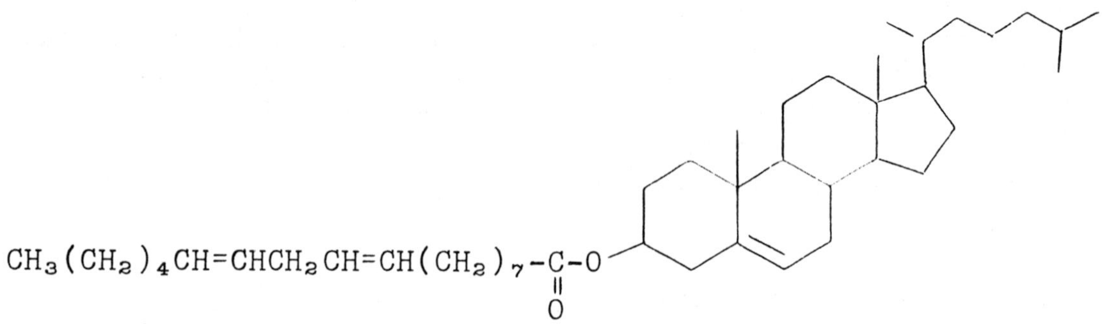

17.30 He or she suffers from hypercholesterolemia which can lead to atherosclerosis and heart attacks.

17.32 It plays a role in the implantation of the fertilized ovum in the uterus.

17.34 (a) They both have a steroid ring structure. (b) RU486 has a <u>para</u>-aminophenyl group on ring C and a triple-bond group on ring D.

17.36 Bile salts are synthesized from cholesterol, so they use up cholesterol and then less is deposited in the form of plaques.

17.38 (a) 3 alcohol groups; amide; carboxylate ion (b) 3 ketone groups; 2 alcohol groups (c) ketone group; 2 alcohol groups; carboxylic acid; 2 C=C double bonds (d) 2 alcohol groups; 4 C=C double bonds; carboxylic acid

17.40 oleic acid: 282; linoleic acid: 280; linolenic acid: 278; arachidonic acid: 304; The unknown acid is linoleic acid.

17.42 $2RCOO^- \; Na^+ + Ca^{2+} \longrightarrow (RCOO^-)_2 \; Ca^{2+} \; (s) + 2Na^+$

17.44 Because soaps form precipitates in hard water; detergents do not.

17.46 the accumulation of glucocerebroside, which enlarges the spleen

17.48 Cholesterol has low solubility in the blood. When present in high levels, it deposits in the form of plaque in the blood vessels, leading to atherosclerosis.

17.50 They prevent ovulation.

17.52 It releases arachidonic acid from complex lipids; this fatty acid forms both prostaglandins and leukotrienes, which cause inflammation.

17.54 It inhibits the synthesis of prostaglandins.

17.56 stearic acid, sphingosine, phosphate, and choline.

17.58 (a) lipases (b) as monoglycerides, alcohols, fatty acids

17.60 175 mg/100 mL or 1.75 g/L

17.62 Taurine is a constituent of bile salts and as such emulsifies lipids and helps enzymes to break them down.

## *Chapter 18. Proteins*

18.2 Most proteins are highly specific in their actions, so a great many are needed to do all the tasks required.

18.4 There is no $NH_2$ group; instead the N is bonded to the R.

18.6 Smallest: glycine Largest: tryptophan

18.8

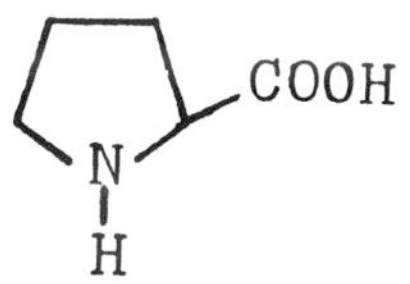

pyrrolidines

18.10 isoleucine, threonine

18.12 Because they are zwitterions. Ionic compounds are solids at room temperature.

18.14

$$CH_3\text{-}CH(NH_3^+)\text{-}COO^- + H_3O^+ \longrightarrow CH_3\text{-}CH(NH_3^+)\text{-}COOH + H_2O$$

$$CH_3\text{-}CH(NH_3^+)\text{-}COO^- + OH^- \longrightarrow CH_3\text{-}CH(NH_2)\text{-}COO^- + H_2O$$

18.16 (a) It is not chiral. (b) The NH is part of a ring.

18.18

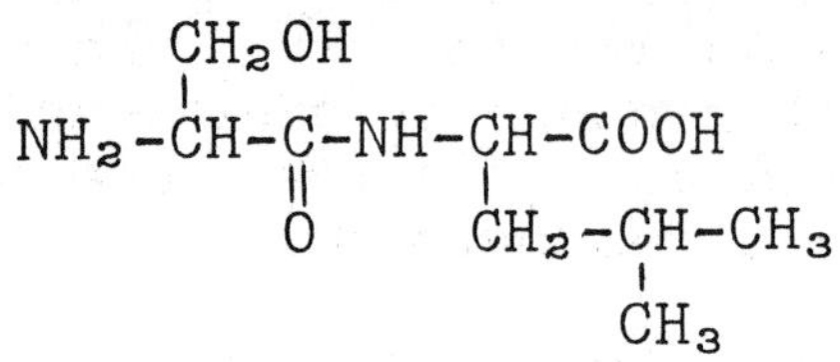

18.20 (a) Val-Phe-Try-Asn (b) C-terminal: Asn
N-terminal: Val

18.22 when it is part of a protein molecule

18.24

at pH 2.0

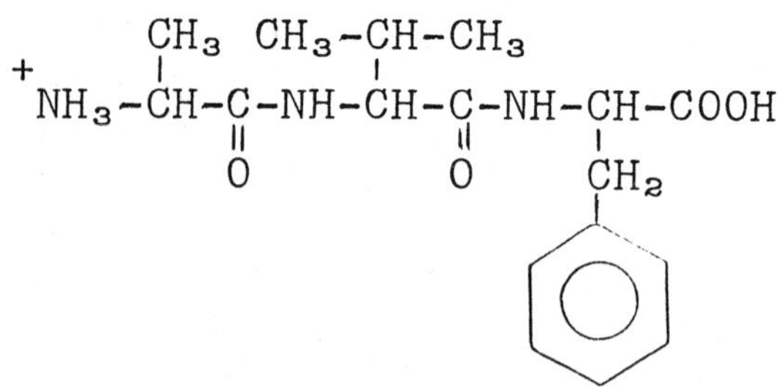

at pH 7.0

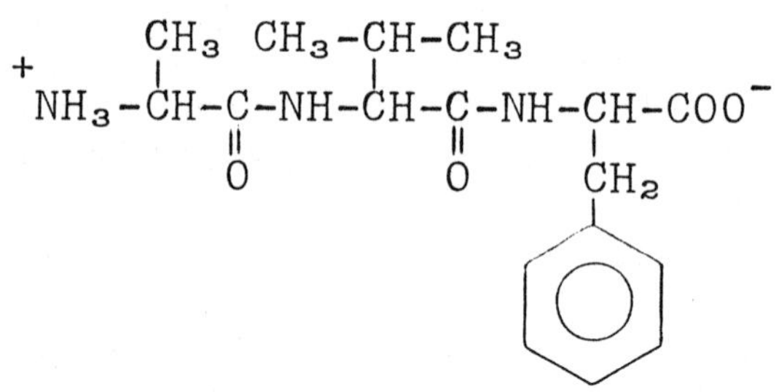

at pH 10.0

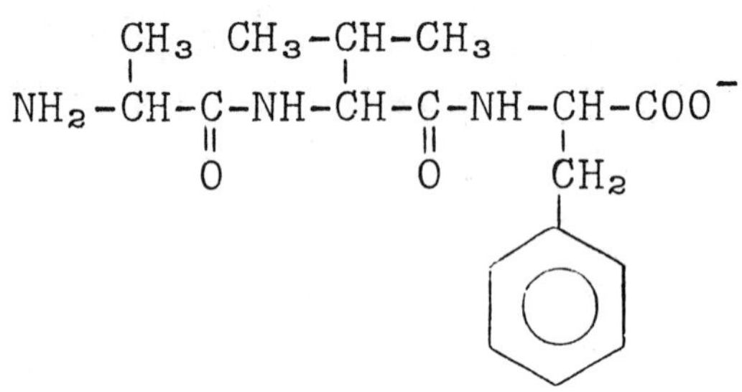

18.26 Proteins contain both positive and negative charges; they can react with acids, taking a proton (the $-COO^-$ becomes $-COOH$) or a base by donating a proton ($-NH_3^+$ becomes $-NH_2$) and the pH will not change.

18.28 (a) 24 (b) $20^4 = 160\ 000$

18.30 (c) most important (b) somewhat less important
(a) least important

18.32 (b)

18.34 An intramolecular bond goes between two parts of the same chain. An intermolecular bond goes between two different molecules. Intermolecular bonds are found in the $\beta$-pleated sheet; intramolecular bonds in the $\alpha$-helix.

18.36 These are nonpolar side chains and go inside the protein to escape from the water molecules outside.

18.38 salt bridge

18.40 a protein that contains a non-amino acid portion called a prosthetic group

18.42 The protein-digesting enzymes in the stomach and intestines would hydrolyze it before it could reach the blood.

18.44 (b)

18.46 Diabetics must restrict their carbohydrate output. To make food taste sweet, they use an artificial sweetener.

18.48 The glutamic acid is linked to the cysteine by its $\gamma$-carboxyl group rather than the $\alpha$-carboxyl group.

18.50 This is unlikely to be detrimental because Asp is similar in structure to Glu (they are both acidic amino acids).

18.52 The gaps in the quarter-stagger arrangement provide the nucleation sites for the hydroxyapatite crystals. The more inorganic the crystal, the harder the tissue: enamel is hardest, bone is softest, and dentine intermediate.

18.54 $Ag^+$ denatures the proteins of bacteria and kills them.

18.56 the particular shape of the keratin molecules of the hair that are stabilized by disulfide linkages.

18.58 Yes. Any change that destroys native proteins is a denaturation.

18.60 It can stabilize 3-dimensional structures if it is oxidized and forms S-S bridges.

18.62 more soluble because carbohydrates have many OH groups that form hydrogen bonds with water

18.64

```
at pH 1        CH3                at pH 12        CH3
               |                                  |
         CH3-CH-CH-COOH                     CH3-CH-CH-COO⁻
                  |                                  |
                  NH3+                               NH2
```

18.66 (a) $4^2 = 16$ (b) $20^2 = 400$

*Chapter 19. Enzymes*

19.2 from $10^9$ to $10^{20}$ times

19.4 It lowers it.

19.6 trypsin

19.8 to join two molecules to form citrate

19.10 (a) deaminations (b) hydrolyses (c) hydrogen removal (d) isomerizations

19.12 Yes. Trypsinogen, a zymogen, is the substrate for some proteases that remove six amino acids from its N-terminal end.

19.14 activation

19.16 No; at high substrate concentration the enzyme surface is saturated and any doubling of the substrate concentration will produce only a slight increase in the rate of the reaction.

19.18 (a) less active at normal body temperature
(b) The activity decreases.

19.20 It is too bulky to fit into the cavity where the active site is located.

19.22 In a competitive inhibition the same maximum rate can be obtained as in the reaction without inhibitor, although at a higher substrate concentration. In a non-competitive inhibition the maximum rate is always lower than that without inhibitor.

19.24 In a multistep enzyme-catalyzed system the accumulation of a substrate at point D can inhibit the catalysis at point A. This is feedback control.

19.26 There is no difference. They are the same.

19.28 1. feedback control 2. negative modulation 3. positive modulation (See Section 19.6.)

19.30 $H_4, H_3M, H_2M_2, HM_3, M_4$

19.32 prostate cancer

19.34 It is denatured by the stomach acid and partially digested. Thus it becomes inactive.

19.36 artifical respiration

19.38 to accomodate the $=CH_2$ group of the PEP

19.40 para-aminobenzoic acid

19.42 the -SH group

19.44 (a) transpeptidase (b) It crosslinks bacterial cell walls.

19.46 Changes in pH, both higher and lower, affect acidic and basic side chains, and consequently salt bridges, which changes conformation of the protein, making it less active.

19.48 Since maximum activity is slightly above pH 7, the enzyme works best in blood.

19.50 They are all basic amino acids.

19.52 hydrolases

19.54 Because dead heart muscles spill their enzyme contents into the serum; unless these enzymes are decomposed they will not appear in the urine.

## *Chapter 20. Bioenergetics. How the Body Converts Food to Energy*

20.2 the citric acid cycle and oxidative phosphorylation

20.4 a fold in the inner membrane of a mitochondrion

20.6 See Figure 20.4.

20.8 adenosine and inorganic phosphate

20.10 No, the storage is only for about a minute.

20.12 (a) dinucleotide (b) There are two phosphate groups attached to the ribose.

20.14 the 2 N atoms that are part of C=N bonds
20.16 ADP
20.18 (a) nicotinamide (b) riboflavin (c) pantothenic acid
20.22 The carbons are converted to $CO_2$; the protons and electrons of the hydrogens are carried away by NADH, $H^+$ and $FADH_2$.
20.24 to form a secondary alcohol (isocitric acid) which can be oxidized, while the tertiary alcohol cannot be
20.26 two
20.28 (a) twice (b) isocitrate, α-ketoglutarate
20.30 (a) succinate (b) oxaloacetate
20.32 No, they originate from the oxaloacetate portion.
20.34 conversion of succinate to fumarate
20.36 to carry electrons from NADH or $FADH_2$ and deliver them to the $O_2$ in the matrix of the mitochondria
20.38 The oxygen is reduced to water by the hydrogen ions and electrons derived from food and produces energy in the form of ATP. The carbon in the food is converted to $CO_2$ during the process.
20.40 at the flavoprotein, Q enzyme and cytochrome $c_1$ stages.
20.42 low, because $H^+$ accumulates there
20.44 Because the energy built into ATP is generated by the osmotic pressure of the accumulated protons across the semipermeable inner membrane of the mitochondrion.
20.46 (a) because it was first made from nicotine (b) because vitamin producers felt that the public would associate nicotinic acid with nicotine
20.48 It uncouples the oxidative process from the phosphorylation by transporting $H^+$ passively, without the expenditure of energy.
20.50 (a) GTP has guanine instead of adenine (b) about the same
20.52 No. The $CO_2$ comes from the alcohol end of the isocitrate, which was originally the oxaloacetate that combined with acetyl CoA.
20.54 (a) 3 (b) 2
20.56 (1) α-Ketoglutaric acid is made by oxidative decarboxylation of isocitric acid. (2) Oxaloacetic acid is made by oxidation of malic acid. (All these are in the form of their salts.)
20.58 108 ATP molecules.

*Chapter 21. Specific Catabolic Pathways: Carbohydrate, Lipid, and Protein Metabolism*

21.2 as their hydrolysis products, fatty acids, glycerol, and monoglycerides
21.4 The glucose must be chemically activated before any subsequent step can take place.
21.6 It occurs in the absence of $O_2$.
21.8 glyceraldehyde 3-phosphate and dihydroxyacetone phosphate
21.10 No; it occurs under aerobic conditions.
21.12 (a) energy-consuming (b) energy yielding

21.14 (a) 4 (b) 0

21.16 (a) 6 (b) 6 (c) 36

21.18 at the dihydroxyacetone phosphate stage

21.20 in the cytosol

21.22 one $CH_3-(CH_2)_{12}-CO-S-CoA$, two $CH_3-CO-S-CoA$, two $FADH_2$, and two $NADH + H^+$

21.24 (a) 5 turns (b) 7 turns

21.26 Thiolase cleaves a $C_2$ fragment from the chain and tranfers it to a CoA molecule; thiokinase converts the RCOOH to RCO-S-CoA.

21.28 fats, because their catabolism produces more ATP molecules than carbohydrates, per carbon atom

21.30 It is a transamination, the first step in catabolism of amino acids.

21.32

$$^{-}OOC-CH_2-\underset{\displaystyle NH_3^+}{\underset{|}{CH}}-COO^- + NAD^+ + H_2O \longrightarrow$$

$$^{-}OOC-CH_2-\underset{\displaystyle O}{\underset{\|}{C}}-COO^- + NADH + H^+ + NH_4^+$$

21.34

$$-\underset{\displaystyle O}{\underset{\|}{C}}-NH_2$$

21.36 (a) 3 (b) 4

21.38 ketone bodies

21.40 It is stored in ferritin and reused.

21.42 the missing or defective enzyme, fructose 1-phosphate aldolase

21.44 It cuts off the oxygen supply, enhancing the glycolysis that produces lactate, which contracts the heart muscle.

21.46 See Box 21D.

21.48 the decarboxylation of the monosodium glutamate ingested

21.50

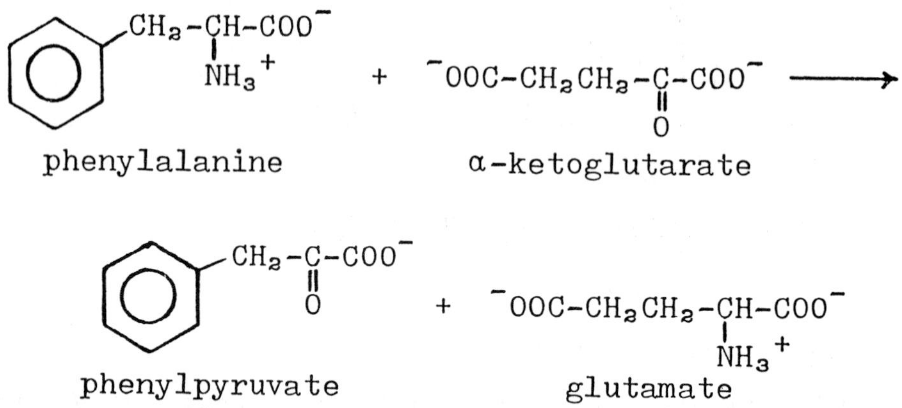


21.52 An incapacitated liver cannot remove bilirubin from the blood.

21.54 diabetes

21.56 2 moles of ATP

21.58 (a) Lysine has one more $CH_2$ group. (b) no

21.60 PEP is able to transfer phosphate to ADP and form one anhydride linkage, producing ATP in step 9 of glycolysis.

## *Chapter 22. Biosynthetic Pathways*

22.2 ATP

22.4 the aerobic metabolism of carbohydrates

22.6 glycolysis

22.8 It has uracil instead of adenine.

22.10 the same

22.12 (a) the number of glucose units in the chain (b) a large number; more than a thousand

22.14 acetyl CoA

22.16 It converts them to fatty acids and then to fats.

22.18 (a) synthase. (b) It picks up a $C_2$ fragment from ACP, combines it with a $C_3$ fragment attached to another ACP, and forms a $C_4$ compound, releasing $CO_2$.

22.20 (c) (e)

22.22

$$\text{ACP-S-(CH}_2)_9\text{-CH}_3 + \text{synthase-SH} \longrightarrow \text{synthase-S-(CH}_2)_9\text{-CH}_3$$

$$\text{synthase-S-(CH}_2)_9\text{-CH}_3 + \text{ACP-S-}\underset{\underset{\text{O}}{\|}}{\text{C}}\text{-CH}_2\text{-COOH}$$

$$\downarrow$$

$$\text{ACP-S-}\underset{\underset{\text{O}}{\|}}{\text{C}}\text{-CH}_2\text{-(CH}_2)_9\text{-CH}_3 + \text{synthase} + \text{CO}_2$$

$$\downarrow \text{NADH} + \text{H}^+ \rightarrow \text{NAD}^+$$

$$\text{ACP-S-}\overset{\overset{\text{H}}{|}}{\underset{\underset{\text{OH}}{|}}{\text{C}}}\text{-CH}_2\text{-(CH}_2)_9\text{-CH}_3 + \text{H}_2\text{O}$$

$$\downarrow$$

$$\text{ACP-S-CH=CH-(CH}_2)_9\text{-CH}_3 + \text{H}_2\text{O}$$

$$\downarrow \text{FADH}_2$$

$$\text{ACP-S-CH}_2\text{CH}_2\text{-(CH}_2)_9\text{-CH}_3$$

22.24 glycerol 3-phosphate, palmitoyl-CoA, lauryl-CoA, and serine

22.26 No; cholesterol is constantly synthesized in the liver.

22.28 oxidative deamination

22.30 $^-\text{OOC-CH}_2\text{-}\underset{\underset{\text{NH}_3^+}{|}}{\text{CH}}\text{-COO}^-$

22.32 (a) chlorophyll (b) in chloroplasts

22.34 (a) indigo blue (b) oxidation of tryptophan
22.36 phenylalanine
22.38 palmitic acid, serine, any other fatty acid (preferably unsaturated), phosphocholine
22.40 malonyl-ACP

*Chapter 23. Nucleic Acids and Protein Synthesis*

23.2 Each gene controls the manufacture of one enzyme.
23.4 histones
23.6 (a) base, sugar, phosphate (b) base, sugar
23.8 See Figure 23.1.
23.10 adenine, cytosine
23.12 A nucleoside contains a sugar and a base. A nucleotide also contains a phosphate
23.14 C-1
23.16 (a)

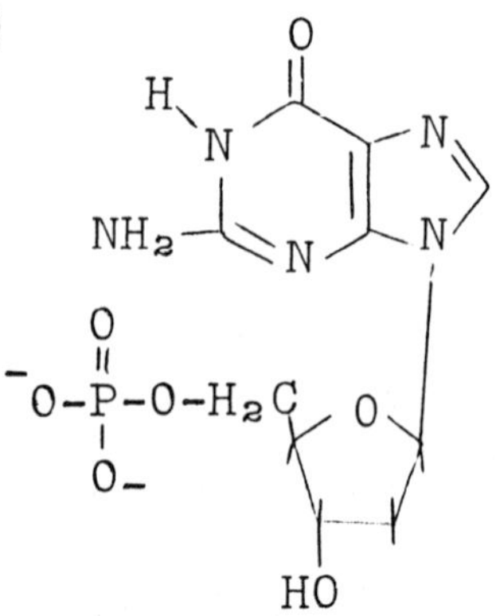

(b)

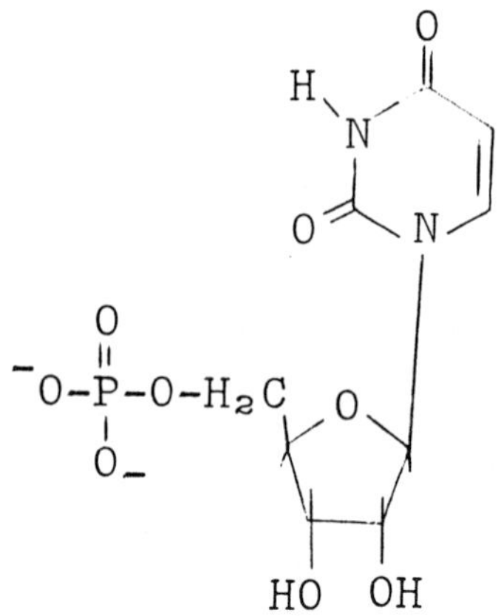

23.18 It allowed the builders of the molecular model to pair A with T and C with G.
23.20 They point inward, to the center.
23.22 Because of hydrogen bonding, only A fits opposite T, and only G fits opposite C.
23.24 See Figure 23.6.
23.26 thousands
23.28 to separate the strands of the double helix by opening a loop or bubble
23.30 from 73 to 93 nucleotides
23.32 The information in DNA is transferred to RNA, and is then expressed in the structure of proteins.
23.34 rRNA
23.36 the 3' end
23.38 (a) the larger (60S) portion (b) See Section 23.10.
23.40 on tRNA
23.42 327 - 2 (for initiation and termination) = 325
23.44 (a) It supports the mRNA. (b) It provides the A and P sites for protein synthesis.

23.46 (a) The introns of DNA are transcribed onto mRNA.
(b) The introns are cut out by enzymes before the mRNA is used for protein synthesis.

23.48 DNA and mRNA

23.50 operator and promoter

23.52 Methylated CG sequences prevent the binding of regulatory proteins to the DNA surface.

23.54 to allow transcription to go much faster by binding to specific groups present in DNA

23.56 (a) Endorestriction nucleases cleave DNA at specific sites. (b) to repair damaged plasmids

23.58

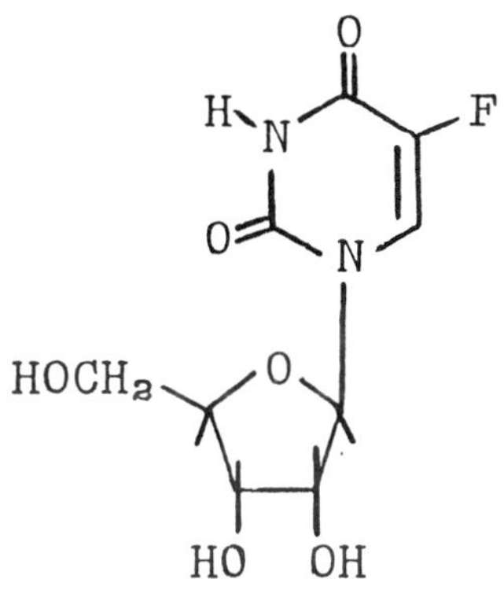


23.60 See Box 23B.

23.62 The enzymes can synthesize DNA only in the 5'⟶3' direction. The leading strand runs this way. The lagging strand runs 3'⟶5', and must be synthesized in short fragments.

23.64 They do not have the molecules necessary to reproduce themselves and must rely on their hosts.

23.66 by animal studies

23.68 The DNA fragments of the child, mother, and the alleged father are run on the same gel and their patterns are compared.

23.70 only by cautioning carriers of defective genes to select noncarrier partners or not to have children

23.72 (a) a certain circular double-stranded DNA molecule
(b) A gene is a double-stranded non-circular DNA molecule that codes for one protein molecule.

23.74 For each amino acid there is a specific enzyme that attaches it to its own specific tRNA.

23.76 See Section 23.2.

## *Chapter 24. Chemical Communication: Neurotransmitters, Hormones, and Immunoglobulins*

24.2 Nerve signals activate endocrine glands; adrenalin is secreted in the blood stream, which makes quick energy available to the muscles.

24.4 (a)(c)(d)(f) See Section 24.2. (b)(e) See Section 24.1.
24.6 Curare inhibits acetylcholinesterase, so that acetylcholine remains at the receptor site and nerve signals can no longer be transmitted.
24.8 the choline end
24.10 C-3 and C-5
24.12 14.6 kcal; ATP $\longrightarrow$ cAMP + $PP_i$
24.14 the same, except that guanine replaces adenine
24.16 The cAMP is destroyed by the enzyme phosphodiesterase.
24.18 (a) an enzyme belonging to the class monoamine oxidase
(b) to oxidize adrenergic neurotransmitters
24.20 By inhibiting MAOs it increases the neurotransmitter concentration The increased concentration provides greater neurotransmitter activity, thus acting as an antidepressant.
24.22 No, the 2 compounds block different receptors.
24.24 Taurine is part of the structure of the bile salt taurocholate.
24.26 They are pentapeptides.
24.28 position 8. A neutral, hydrophobic side chain, Leu, in oxytocin is replaced by a basic side chain, Arg, in vasopressin.
24.30 (a)(b) small molecules derived from amino acids
(c) peptides (d) steroids
24.32 IgG
24.34 the Fab fragments; they carry the variable regions that bind the antigen.
24.36 Many communication molecules, such as cell adhesion molecules, have structures similar to that of immunoglobulins. They form a superfamily of compounds.
24.38 intermolecular attractions: London dispersion forces, dipole-dipole interactions, and hydrogen bonds
24.40 It is converted to acetyl CoA.
24.42 It blocks $Na^+$ ion channels in nerve membranes, so nerve transmissions cannot take place.
24.44 See Box 24D.
24.46 Insulin-dependent patients do not manufacture enough insulin. Non-insulin dependent patients have enough insulin, but not enough receptors to which insulin can bind.
24.48 See Box 24G.
24.50 They kill invading cells when they touch them.
24.52 See Box 24H.
24.54 the IgA immunoglobulins
24.56 Yes, decamethonium is a competitive inhibitor of acetylcholinesterase.
24.58 They have variable regions that react with antigens.

## Chapter 25. Body Fluids

25.2 interstitial fluid
25.4 erythrocytes, leucocytes, platelets or thrombocytes
25.6 plasma, serum, urine
25.8 (a)(b) in the bone marrow (c) in the lymph nodes (d) in the bone marrow and spleen
25.10 (a) to provide proper osmotic pressure (b) to provide immune reactions (c) blood clotting
25.12 See Section 25.2.
25.14 660 mm Hg
25.16 4 = 100 mm Hg; 3 = 36 mm Hg; 2 = 27 mm Hg; 1 = 18 mm Hg; 0 = 0 mm Hg
25.18 (a) 4 (b) 4
25.20 (a) It decreases. (b) Bohr effect
25.22 $\underline{K}$ = 9.9
25.24 bicarbonate/carbonic acid
25.26 the distal tubule and the collecting tubule
25.28 See Section 25.8.
25.30 No glucose reaches the tissues or the cells. Because 12 mm Hg is lower than the osmotic pressure, nutrients flow from the interstitial fluid into the blood.
25.32 drowsiness
25.34 platelets
25.36 (a) Blood donors are not needed; blood types need not be matched; the blood substitute can be stored longer. (b) It performs only 1 of the functions of whole blood.
25.38 The Fluosol-DA particles carrying $O_2$ are smaller than the red blood cells. They can pass by partial blockage and deliver $O_2$ to the heart muscle, preventing its death.
25.40 (a) blood plasma, interstitial fluid (b) intracellular fluid
25.42 It blocks adrenoreceptor sites. See Box 25E.
25.44 See Box 25E.
25.46 The concentration is higher. Since fibrinogen is removed when plasma becomes serum, the albumin is dissolved in a smaller amount of material, and its concentration is higher.
25.48 vasopressin
25.50 Less oxygen is carried; more is released to the tissues.

## Chapter 26. Nutrition and Digestion

26.2 (a) no (b) It is a carboxylic ester.
26.4 for a young adult male, 16.4 ounces, or 465 g; for a female 11.8 ounces, or 335 g
26.6 (a) breakfast cereal (b) tuna
26.8 for daily activities involving motion
26.10 hypertension, cardiovascular disease, and diabetes
26.12 fats

26.14 (a) 2 (b) 10
26.16 Add animal proteins to the diet.
26.18 No; amylose does not contain branches.
26.20 night blindness
26.22 (a) vitamin $B_{12}$ (b) part of methyl-removing enzyme
26.24 meat and dairy products
26.26 degradation of the nervous system
26.28 We have enzymes that catalyze the hydrolysis of $\alpha$ linkages, but not $\beta$ linkages.
26.30 Trypsin hydrolyzes proteins only at an arginine or lysine residue; HCl does it randomly.
26.32 If glucose supplied all the caloric needs, it would make the blood hypertonic and cause sclerosis and deterioration of peripheral veins.
26.34 Eat some candy or other sugar-containing food.
26.36 (a) 15 mg (b) in enzymes, as a cofactor, and in insulin
26.38 They contain the essential amino acids we cannot synthesize ourselves.
26.40 Yes, but one must eat a wide range of vegetables and cereals.